£3

CAMBRIDGE MONOGRAPHS ON PHYSICS

GENERAL EDITORS

M. M. WOOLFSON, D.Sc.
Professor of Theoretical Physics, University of York
J. M. ZIMAN, D.PHIL., F.R.S.
Henry Overton Wills Professor of Physics, University of Bristol

Pulsars

Pulsars

F. G. SMITH, F.R.S.
Director, Royal Greenwich Observatory, Herstmonceux

CAMBRIDGE UNIVERSITY PRESS
CAMBRIDGE
LONDON · NEW YORK · MELBOURNE

Published by the Syndics of the Cambridge University Press
The Pitt Building, Trumpington Street, Cambridge CB2 1RP
Bentley House, 200 Euston Road, London NW1 2DB
32 East 57th Street, New York, NY 10022, USA
296 Beaconsfield Parade, Middle Park, Melbourne 3206, Australia

© Cambridge University Press 1977

First published 1977
Reprinted with corrections 1979

Printed in Great Britain at the University Press, Cambridge and set on Linotron Filmsetter by J. W. Arrowsmith Ltd, Bristol, England.

Library of Congress Cataloguing in Publication Data

Smith, Francis Graham, 1923–
Pulsars.

(Cambridge monographs on physics)
Includes index.
1. Pulsars.
QB843.P8S56   523   75-44569
ISBN 0 521 21241 3

Contents

Preface — xi
Some astronomical quantities — xii

1 The discovery of the pulsars — 1
  1.1 Interplanetary scintillation — 2
  1.2 The *Nature* letter of February 1968 — 3
  1.3 The identification with neutron stars — 4
  1.4 Optical pulses from the Crab Pulsar — 6
  1.5 X-ray pulses from the Crab Pulsar — 8
  1.6 The development of pulsar research — 8

2 Techniques for search and for observation — 11
  2.1 The problem of sensitivity — 11
  2.2 Frequency dispersion in pulse arrival time — 12
  2.3 Search techniques: single pulses — 13
  2.4 Search techniques: periodicity — 15
  2.5 Optical, X-ray and gamma-ray techniques — 16

3 The identification with rotating neutron stars — 19
  3.1 Oscillations of condensed stars — 19
  3.2 Planetary and binary orbits — 21
  3.3 Rotating condensed stars — 22
  3.4 Changes in period — 23

4 The X-ray pulsars — 25
  4.1 Her X-1/HZ Her — 26
  4.2 Cen X-3 — 29
  4.3 Period changes and mass transfer — 30
  4.4 Other known X-ray binaries — 32
  4.5 The physics of the X-ray emission — 33

5 The internal structure of neutron stars — 36
  5.1 Temperature — 36
  5.2 The outer layers — 38
  5.3 The neutron drip point — 39
  5.4 The neutron fluid: superfluidity — 40

Contents

5.5	Solidification of the core and formation of hyperons	40
5.6	Neutron star models	41
5.7	The stability and the formation of neutron stars	42

6 The magnetosphere of neutron stars — 46
6.1	Electrodynamics of the magnetosphere	46
6.2	Axisymmetric dipole	48
6.3	The oblique rotator	51
6.4	Complications and instabilities	51
6.5	Ruderman's whiskers and the work function	52
6.6	Energy flow from the pulsar	53

7 Pulse timing — 55
7.1	Pulsar positions and the barycentric correction	55
7.2	The relativistic correction	58
7.3	Periods, period changes and pulsar positions	59
7.4	The Crab Pulsar	60
7.5	The Vela Pulsar PSR 0833-45	64
7.6	Neutron star structure and the glitch function	65
7.7	Starquakes and corequakes	67
7.8	The age of pulsars and the braking index	68
7.9	Pulsar proper motion	70
7.10	Binary systems	71
7.11	The binary pulsar PSR 1913 + 16	73

8 Properties of the integrated radio pulses — 76
8.1	Integrated pulse profiles	76
8.2	On/off ratio : interpulses	80
8.3	Frequency dependence of the integrated pulse profiles	81
8.4	Integrated polarisation	82
8.5	Fine structure in integrated profiles	85
8.6	Mode changing	85
8.7	The spectrum of pulsar radio emission	88
8.8	Long-term variations of flux density	88

9 Individual radio pulses — 91
9.1	The structure of individual pulses	92
9.2	Width of the sub-pulses	96
9.3	Microstructure	100
9.4	Histograms of pulse energy	101
9.5	Pulse drifting	103
9.6	Modulation by pulse drifting	105
9.7	Nulling and moding	106

10 The Crab Nebula — 109
- 10.1 Discovery and early observations — 109
- 10.2 The continuum radiation from the Crab Nebula — 111
- 10.3 The energy supply — 112
- 10.4 The transfer of energy from the pulsar to the nebula — 114

11 The Crab Pulsar — 117
- 11.1 The spectrum — 117
- 11.2 Pulse shapes — 120
- 11.3 Dispersion measure and pulse arrival times — 121
- 11.4 Variations of intensity — 122
- 11.5 Variability of pulses: giant pulses — 122
- 11.6 Polarisation: optical — 124
- 11.7 Polarisation: radio — 126
- 11.8 Relation between optical and radio radiation from the pulsar — 127

12 The interstellar medium as an indicator of pulsar distances — 129
- 12.1 The interstellar electrons — 129
- 12.2 The galactic HI disc — 130
- 12.3 The disc of HII regions — 132
- 12.4 A model of the electron distribution — 134
- 12.5 Absorption in neutral hydrogen — 134
- 12.6 Pulsars at high galactic latitudes — 137

13 The interstellar magnetic field — 140
- 13.1 Eighteen orders of magnitude down — 140
- 13.2 Faraday rotation in pulsars — 142
- 13.3 The configuration of the local field — 143
- 13.4 The more distant field — 145
- 13.5 The galactic magnetic field: summary — 146

14 Interstellar scintillation — 148
- 14.1 A thin screen model — 148
- 14.2 Diffraction theory of scintillation — 150
- 14.3 Thick (extended) scattering screens — 152
- 14.4 Observational results — 153
- 14.5 Pulse broadening — 156
- 14.6 Multiple scattering: geometric approach — 158
- 14.7 Diffraction theory of pulse broadening — 160
- 14.8 Observations of pulse broadening — 162
- 14.9 Apparent source diameters — 165
- 14.10 The velocity of the scintillation pattern — 166
- 14.11 Proper motions of the pulsars — 169

Contents

15	**Radiation processes**	171
15.1	Cyclotron radiation	171
15.2	Cyclotron radiation from streaming electrons	173
15.3	Synchrotron radiation	174
15.4	Curvature radiation	177
15.5	Coherence	178
15.6	Maser amplification	179
16	**The emission mechanism I: analysis of observed properties**	180
16.1	The integrated pulse profiles	180
16.2	Individual pulses: the sub-pulses	181
16.3	Intensity and spectrum	182
16.4	The period P and its derivative $\dot{P}$	183
16.5	Radiation characteristics related to age and period	184
17	**The emission mechanism II: geometrical considerations**	186
17.1	The width of the integrated profile	186
17.2	The shape of the integrated profile	190
17.3	Sub-pulse width	190
17.4	Relativistic beaming	190
17.5	Sub-pulse polarisation	195
17.6	Position angle of polarisation	196
17.7	Single vector model	197
17.8	Relation to a rotating magnetic field	198
17.9	Conclusions: the location of the emitter	199
18	**The emission mechanism: discussion**	201
18.1	Location	201
18.2	Energy density	204
18.3	Spectrum	205
18.4	Radio frequency radiation	206
18.5	Concluding discussion	208
19	**Supernovae: the origin of the pulsars**	211
19.1	Supernova explosions	211
19.2	Frequency of occurrence of supernovae	213
19.3	Supernovae in binary systems	214
19.4	The chances of observing a binary pulsar	217
19.5	Velocities acquired by neutron stars	218
19.6	Associations between pulsars and supernovae	218
20	**The distribution and the ages of pulsars**	220
20.1	The observed and actual distribution functions	221
20.2	Distribution in period	222

20.3	Distribution in luminosity	222
20.4	Distribution in z-distance	223
20.5	Distribution with radial distance R	223
20.6	The total galactic population	224
20.7	Correlations between the population functions	224
20.8	The ages of the pulsars	225
20.9	The radiation cut-off	227
20.10	Conclusions	228

21 High energies and condensed stars — 230

21.1	X-ray binaries	230
21.2	X-rays in clusters	231
21.3	The galactic centre	232
21.4	Extragalactic clusters	232
21.5	Radio galaxies and quasars	233

Appendix: the positions and periods of 105 pulsars — 234

Index — 237

Preface

The discovery of the pulsars by Professor Hewish and Miss Jocelyn Bell in 1967 ranks with the discovery of quasars and of the universal microwave background radiation as one of the major advances in modern astronomy. At that time the techniques of radio astronomy were sufficiently well developed in several observatories for rapid advances to be made in the discovery of further pulsars and their properties, while theorists quickly grasped the significance of the discoveries, producing interpretations at a bewildering rate and, it must be admitted, in bewildering variety. It soon became difficult for most astronomers, and almost impossible for other scientists, to follow the pulsar story beyond the simple association between pulsars, neutron stars, and supernova explosions. The pulsar observers themselves were trying rather unsuccessfully to find critical questions which their data could enlighten, while the theorists were analysing models which were obviously inadequate oversimplifications. The result was a flood of complicated and confusing publication.

This flood of papers has now subsided, and it is time to review the observations and theory of pulsars. For the author 1975 is an appropriate time, after an exciting 6 years of pulsar work with his colleagues at Jodrell Bank. The work began immediately the discovery was announced, because of the opportunities provided by the 250-foot radio telescope and excellent on-line computer system developed for it by Professor J. G. Davies. The research at Jodrell Bank continues under the leadership of my colleague Dr A. G. Lyne, to whom I am greatly indebted. I myself have now moved from radio to optical astronomy, where observing possibilities in the pulsar field are very limited, and I can contribute to the subject mainly through this review. It is to be hoped that it will be useful to a wider range of scientists than the small band of pulsar specialists, who will recognise that it solves no problems, that it is biassed towards a description of phenomena rather than to abstruse theory, and above all that it depends almost entirely on their own published work and very little on my own. I hope they will forgive me for the particular selection of work which I have chosen.

Some Astronomical Quantities

Astronomical unit (AU)	$= 1.496 \times 10^{13}$ cm
Parsec (pc)	$= 3.085 \times 10^{18}$ cm
	$= 3.26$ light years
Tesla (T)	$= 10^4$ gauss
Earth equatorial radius	$= 6378$ km
Sun – radius	$= 6.96 \times 10^5$ km
– mass	$= 1.989 \times 10^{33}$ g
Electron – mass	$= 9.108 \times 10^{-28}$ g
Proton – mass	$= 1.672 \times 10^{-24}$ g
Electron charge	$= 1.602 \times 10^{-19}$ coulomb
Electron volt (eV)	$= 1.602 \times 10^{-19}$ joule
Equivalent wavelength (1 eV)	$= 1240$ nm

Pulse delay t: ν is frequency (Hz), n_e is electron density (cm^{-3}), L is distance (cm)

$$t = 1.345 \times 10^{-3} \nu^2 n_e L \text{ s}.$$

Pulse delay t: ν is frequency (MHz), DM is dispersion measure (pc cm^{-3})

$$t = 4.15(DM)\nu^{-2} \text{ s}.$$

Faraday rotation angle θ(radians): n_e is electron density (cm^{-3}), H is parallel component of magnetic field (μG), l is distance (pc), λ is wavelength in metres, R is rotation measure:

$$\theta = 0.18\lambda^2 \int n_e H \, dl = R\lambda^2.$$

1
The discovery of the pulsars

The spectacular growth of radio astronomy during the 30 years following the Second World War was marked by the introduction of a series of new observational techniques, each of which opened new observational fields of research. Each advance in technique was applied initially to a specific problem, but such was the richness of the radio sky that each new technique guided the observers into unexpected directions. Examples of such serendipity are provided by all the major discoveries in radio astronomy. An investigation of the radio background, undertaken by J. S. Hey and his colleagues as an extension of their meteor radar work, yielded as a complete surprise the first discrete radio source, Cygnus A. New techniques in radio telescopes, designed in Cambridge to extend the counts of extragalactic sources for cosmological investigations, led unexpectedly to the discovery of quasars; the identification of quasars itself depended on the development of lunar occultation and its use with the new radio telescope in Sydney. The added bonus, which so often follows an adventure into a new observational technique, has its example *par excellence* in the discovery of the pulsars.

At the start of the story we may ask why it was that pulsars were not discovered earlier than 1967. Their signals are very distinctive and often quite strong, so that, for example, the 250-foot radio telescope at Jodrell can be used to produce audible trains of pulses from several pulsars. The possibility of discovery had existed for 10 years before it became reality. In fact, it turned out that pulsar signals had been recorded but not recognised when this telescope was used for a survey of background radiation several years before the actual discovery. The pulsar now known as PSR 0329 + 54 left a clear imprint of strong pulses on several of the survey recordings. A similar story can be told for radio pulses from the planet Jupiter: these were discovered in 1954, although recordings made in Australia 5 years previously contained signals from Jupiter which proved to be useful in subsequent analyses of the rotation period of the radio sources on the planet. An even more remarkable pre-discovery recording exists for the X-ray pulses from the Crab Pulsar: these were found on records from a balloon flight which pre-dated the technical *tour-de-force* of the actual discovery rocket flights by 2 years. The signals were recorded, but not recognised.

The initial difficulty in recognition of the pulsar radio signals was that radio astronomers were not expecting to find rapid fluctuations in the signals from any celestial source. An impulsive radio signal received by a radio telescope was regarded as interference, generated in the multitudinous terrestrial impulsive sources, such as electrical machinery, power line discharges, and automobile ignition. Indeed, most radio receivers were designed to reject or smooth out impulsive signals and to measure only steady signals, averaged over several seconds of integration time. Even if a shorter integrating time constant was in use, a series of impulses appearing on a chart recorder would excite no comment; interference of such regular appearance is to be expected, and is often encountered from such a simple device as an electric fence on a farm within a mile or two of the radio telescope.

Two attributes were lacking in the apparatus used in these previous surveys: a short response time and a repetitive observing routine, which would show that the apparently sporadic signals were in fact from a permanent celestial source. These were both features of the survey of radio scintillation designed by A. Hewish, in the course of which the first pulsar was discovered.

1.1 Interplanetary scintillation

The familiar twinkling of visible stars, due to random refraction in the terrestrial atmosphere, has three distinct manifestations in radio astronomy, in which the refraction is due to ionised gas. Radio waves are refracted in neutral air in much the same way as light, but there are much larger refraction effects for radio waves in ionised gas. Random refraction, causing scintillation of radio waves from celestial sources, occurs in the terrestrial ionosphere, in the ionised interplanetary gas in the solar system, and in the ionised interstellar gas of the Galaxy. In all three regions the radio waves from a distant point source traverse a medium with fluctuations of refractive index sufficient to deviate radio rays into paths which cross before they reach the observer, giving rise to interference. All three types of radio twinkling, or 'scintillation', were discovered and investigated at Cambridge, and in all three investigations Hewish played a key part. Coincidentally, the theory of interstellar scintillation is important for pulsars, which now provide its most dramatic demonstration; the coincidence is that it was an investigation of interstellar scintillation that led to the discovery of pulsars, even though the discovery was a by-product rather than the purpose of the investigation.

Hewish was working with a research student, Miss Jocelyn Bell (now

1.2 The Nature *letter of February 1968*

Dr Burnell). They constructed a large receiving antenna for a comparatively long radio wavelength, 3.7 m, making a radio telescope which was sensitive to weak discrete radio sources. At this long wavelength the interstellar scintillation effects are large, but they only occur for radio sources with a very small angular diameter. Scintillation is therefore seen as a distinguishing mark of the quasars, since the larger radio galaxies do not scintillate; Hewish later used the results of a survey with this system to study the distribution and population of these very distant extragalactic sources. The observational technique involved a repeated survey of the sky, using a receiver with a short time constant which would follow the scintillation fluctuations, these being rapid for long radio wavelengths.

The discovery was made by Jocelyn Bell within a month of the start of regular recordings in July 1967. Large fluctuations of signal were seen at a time repeating for several days, as would a signal of celestial origin. The characteristics of the signal looked unlike scintillation, and very like terrestrial interference. Hewish at first dismissed them as interference, such as might be picked up from a passing motor car. For several nights no signals appeared; as we now know, this must have been due to the chance scintillation effects. Then they re-appeared, and continued to re-appear spasmodically until in October Hewish concluded that something new had turned up. What sort of celestial source could this be? He and his colleagues then used a recorder with an even faster response time, and in November they first saw the amazingly regular pulses. Could they be man-made? Possibly they originated on a space-craft? Possibly they were the first radio signals from an extraterrestrial civilisation?

The last possibility was disturbing. If it became known to the public that signals were being received which might have come from intelligent extraterrestrial sources, the 'little green men' of space fiction, the newspaper reporters would descend in strength on the observatory and destroy any chance of a peaceful solution to the problem. So there was intense activity, but no communication for two months, until in February 1968 a classic paper appeared in *Nature* (Hewish *et al.*, 1968).

1.2 The *Nature* letter of February 1968

The announcement of the discovery contained a remarkable analysis of the pulsating signal, which already showed that the source must lie outside the Solar System, and probably at typical stellar distances; furthermore, it showed that the source must be some form of condensed star, presumably either a white dwarf or a neutron star. These results have, of course, been expanded and overtaken by later work, but it should not be forgotten that in the completely open and speculative atmosphere of the first few weeks after the discovery the right conclusions were

The discovery of the pulsars

reached on the most important properties of the pulsating source. The location outside the Solar System came from observations of the effect of the Earth's motion on the pulse periodicity: this phenomenon also led to a position determination. Admittedly, the first observations miscounted by one the number of pulses in one day, giving an error in period of 1 part in 6×10^4; but this seems to be the only subsequent correction to the pulse timing. It is particularly interesting to see that the paper specifically mentions a neutron star as a possible origin, when at that time the existence of neutron stars was only hypothetical. Indeed, the flood of speculative theoretical papers which was let loose by the discovery did not even follow up this idea at first, exploring instead every possible configuration of the more familiar binary systems and white dwarf stars.

A few days before the *Nature* letter appeared, the discovery was discussed at a colloquium in Cambridge. The news spread rapidly, and radio astronomers immediately turned their attention to confirming the remarkable results. Only a fortnight separated the first paper and a *Nature* letter from Jodrell Bank giving some remarkable extra details of the radio pulses from this first pulsar, now known as CP 1919. New discoveries of pulsars were made and announced by other observatories within a few months. By the middle of the year significant contributions were being made by at least eight radio observatories.

This celebration of the discovery of pulsars was somewhat marred by allegations that Cambridge had withheld publication of its results instead of making all information freely available at the moment of discovery. The trouble seems to have been caused by a statement in the original *Nature* announcement that three other pulsars had been detected, and that their characteristics were being investigated. The statement was evidently made merely as supporting evidence for the astrophysical interpretations advanced in the *Nature* letter, but it led to a bombardment of requests for advance information on the location and periodicities of these three further pulsars. Hewish refused to give further details until his measurements were complete; the results were published in *Nature* in April. His action was entirely in accordance with normal scientific protocol, but it was misinterpreted. Unfortunately some lingering hurt feelings can still be detected more than 6 years later; perhaps some student of the sociology of science may be able to unravel the tangled skeins of jealousy and pride which seem to have been at work.

1.3 The identification with neutron stars

The historian of science will also enjoy the story of the theoretical papers which led to the identification of pulsars with neutron stars. It should be

1.3 The identification with neutron stars

remembered that white dwarf stars were already observable and understood, while the further stage of condensation represented by a neutron star existed in a theory familiar only to certain astrophysicists who were concerned with the solid state of matter. Suggestions based on the more familiar white dwarf stars, and particularly on their various possible modes of oscillation, poured out from the theorists. Unknown to them, and apparently also unnoticed by Hewish, Dr Franco Pacini had already published a paper containing the solution to the nature of pulsars, again in *Nature*, and only a few months before the discovery. This paper (Pacini, 1967) showed that a rapidly rotating neutron star, with a strong dipolar magnetic field, would act as a very energetic electric generator, which could provide a source of energy for radiation from a surrounding nebula, such as the Crab Nebula.

In June 1968 *Nature* published a letter from Professor T. Gold, of Cornell University, which set out very clearly the case for identifying the pulsars with rotating neutron stars. Between them, the two papers from Pacini and Gold contained the basic theory and the vital connection with the observations. The remarkable part of the story is that the two men were working in offices practically next door to one another at the time of Gold's paper, since Pacini was visiting Cornell University; nevertheless Gold did not even know of Pacini's earlier work, and there is no reference to it in his paper (Gold, 1968). Collaboration was, of course, soon established, as may be seen from a paper from Pacini only a month later (Pacini, 1968). These two men should clearly share the credit for the linkage between pulsars and neutron stars.

The confusion of theories persisted until the end of 1968, even though the correct theory had been clearly presented. Unfamiliarity with the concept of a neutron star seems to have been the main barrier to understanding, at least for the observers; it is interesting to see that both Hewish and the present author wrote forewords to a collection of *Nature* papers towards the end of 1968 in which they favoured explanations involving the more conventional white dwarf stars. The issue was settled dramatically by the discoveries of the short-period pulsars now known as the Vela and the Crab Pulsars. The experimental test was simple: theories involving white dwarf stars might account for pulsars with periods of about 1 s, and possibly even for $\frac{1}{4}$ s, the shortest period then known, but the Vela Pulsar, discovered in Australia by Large, Vaughan & Mills (1968), had a period of only 88 ms, while the Crab Pulsar, discovered in USA by Staelin & Reifenstein (1968) had the even shorter period of 33 ms. Only a neutron star could vibrate or rotate as fast as 30 times per second. Furthermore, as pointed out by Pacini and by Gold, a rotation

The discovery of the pulsars

would slow down, but a vibration would not. Very soon a slowdown was discovered in the period of the Crab Pulsar (Richards & Comella, 1969), and the identification with a rotating neutron star was then completely certain.

The astrophysical arguments leading to the identification are presented in more detail in Chapter 3.

1.4 Optical pulses from the Crab Pulsar

The possibility that pulsars might emit pulses of light as well as radio was tested on the first pulsar, CP 1919, as early as May 1968. In the excitement some over-optimistic positive results were reported at first both from Kitt Peak and Lick Observatories, but eventually every attempt was abandoned without any detection of optical pulsation or variation of any kind in several radio pulsars. Photometric equipment was, however, assembled for searches for periodic fluctuations in white dwarf stars, and on 24 November 1968 a first recording of the centre of the Crab Nebula was made by Willstrop in Cambridge without prior knowledge of the discovery of the radio pulsar a few days earlier in the USA (Willstrop, 1969). Although this recording subsequently was found to show the optical pulsations of the Crab Pulsar, it was stacked away with others for a subsequent computer analysis, and the discovery went instead to an enterprising team at the Steward Observatory in Arizona who were among three groups of observers fired with enthusiasm by the discovery of the Crab Pulsar.

The discovery of the optical pulses by Cocke, Disney & Taylor is properly recorded in a *Nature* letter (Cocke *et al.*, 1969); less usually, the actual event of the discovery was recorded on a tape recorder, which was accidentally left running at the time. The excitement of the appearance of a pulse on a cathode ray tube, after a few minutes of integration, is well conveyed by the uninhibited (and unprintable) remarks of the observers. The discovery was made on 16 January 1969. Only three nights later the light pulses were observed by two other groups, at McDonald Observatory and at Kitt Peak Observatory. Shortly afterwards a new television technique was applied to the 120-inch reflector at Lick Observatory, and a stroboscopic photograph of the pulsar was obtained. This showed two contrasting exposures, made at pulse maximum and minimum (Fig. 1.1; Miller & Wampler, 1969).

Fig. 1.1. The Crab Pulsar. This pair of photographs was taken by a stroboscopic technique, showing the pulsar on (*above*) and off (*below*). (Lick Observatory, reproduced by kind permission of the Royal Astronomical Society Library.)

1.4 Optical pulses from the Crab Pulsar

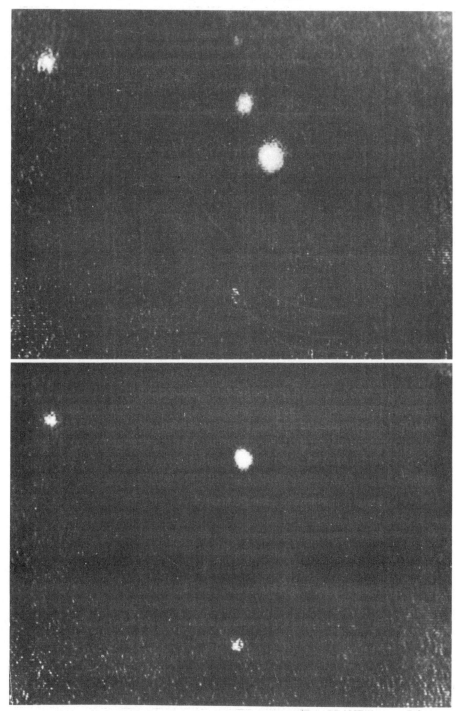

The discovery of the pulsars

Subsequent observations have, of course, given very much more detail about the pulse timing, pulse shape, spectrum, and polarisation of these optical pulses; as might be expected, these are recorded in less dramatic form than the first paper by Cocke, Disney & Taylor, and their accidental, historic tape recording.

1.5 X-ray pulses from the Crab Pulsar

The final link in the chain of discoveries about the Crab Pulsar was the extension of the spectrum into the X-ray region. The observations were necessarily made from above the Earth's atmosphere. In 1969 there was no X-ray telescope orbiting the Earth in a satellite, so that the only possibility lay in rocket flights. Astonishingly, two such rocket flights were successfully made, within a week of one another, and only three months after the discovery of optical pulses. The first was made by a team from the Naval Research Laboratory Washington (Fritz *et al.*, 1969), and the second from the Massachusetts Institute of Technology (Bradt *et al.*, 1969).

Both were completely successful, showing that the pulsed radiation extended to X-ray energies of several kilovolts: in fact, the total power radiated in the X-ray region was found to be at least 100 times that in visible light. The shape of the pulses was very nearly the same in X-rays as in light.

The Crab Nebula had been known and studied for several years as a source of X-rays. After the two rocket flights designed especially for the detection of periodic pulses had demonstrated the existence of the pulsar within the nebula, the recordings of an earlier rocket flight were re-examined; they showed that the pulses had been recorded but not recognised. This flight was in March 1968 (Boldt *et al.*, 1969). Even this pre-discovery recording turned out not to be the earliest, since a balloon-borne experiment in 1967 designed to measure the spectrum of the Crab Nebula up to X-ray energies of 20 keV was found to have recorded the periodic 'light curve' of the pulsar (Fishman, Harnden & Haymes, 1969). There was sufficient accuracy in the periods obtainable from these two earlier experiments to show that the pulsar had been slowing down at the same average rate prior to the discovery as afterwards.

1.6 The development of pulsar research

The richness of the new field of research was demonstrated by a very rapid branching of the subject matter, even during 1968. There was, of course, from the start a division between theoretical and experimental papers, but the theories and observations arrived chaotically, without the

References

dignified alternation of hypothesis and experimental test that is supposed traditionally to represent the progress of science. When the dust had cleared from this initial explosion, the theorists were seen to be at work on several lines: the solid-state theory of the interior of neutron stars, the crystalline state of the surface, the magnetosphere, and the radiation mechanism which produced the pulses. The observers were ordering their work towards the determination of periods and positions, the organised search for more pulsars, the description of physical conditions in the pulse emitter, and the use of the pulses in exploring interstellar space. The results of the first seven years of these researches form the subject matter of this book.

The following chapters will show that considerable success has been achieved in the understanding of the interior of the neutron star, but that little is known about the magnetosphere and the mechanism of the radiation. The population of pulsars within the Galaxy is now known to fit well with the hypothesis that the neutron stars are created in supernova explosions, but the detail of this creation is not understood. The exploration of interstellar space has led to a measured value of the interstellar magnetic field, and, through scintillation experiments, to a demonstration of very high velocities in the pulsars: the origin of these high velocities is not understood.

The spectrum of phenomena has also been broadened by the discovery of the X-ray pulsars, which are also neutron stars but whose X-radiation depends on their membership of binary systems. The physics of these sources is different from that of the general run of pulsars which are described in this book. At first it was possible to distinguish the latter as the solitary pulsars, since none was a member of a binary system. There is now one pulsar which is a member of a close binary system, in which the clock-like precision of the pulses provides a marvellous test of relativity theory.

One exciting aspiration which has not yet been achieved is the detection of a pulsar in an extragalactic system. Perhaps this will be included in the historical introduction to a later book on pulsars.

References

Boldt, E. A., Desai, U. D., Holt, S. S., Serlemitsos, P. J. & Silverberg, R. F. (1969). *Nature, Lond.* **223**, 280.
Bradt, H., Rappaport, S., Mayer, W., Nather, R. E., Warner, B., Macfarlane, M. & Kristian, J. (1969). *Nature, Lond.* **222**, 728.
Cocke, W. J., Disney, M. J. & Taylor, D. J. (1969). *Nature, Lond.* **221**, 525.
Fishman, G. J., Harnden, F. R. & Haymes, R. C. (1969). *Astrophys. J.* **156**, L107.

Fritz, G., Henry, R. C., Meekins, J. F., Chubb, T. A. & Friedman, H. (1969). *Science* **164**, 709.
Gold, T. (1968). *Nature, Lond.* **218**, 731.
Hewish, A., Bell, S. J., Pilkington, J. D. H., Scott, P. F. & Collins, R. A. (1968). *Nature, Lond.* **217**, 709.
Large, M. I., Vaughan, A. F. & Mills, B. Y. (1968). *Nature, Lond.* **220**, 340.
Miller, J. S. & Wampler, E. J. (1969). *Nature, Lond.* **221**, 1037.
Pacini, F. (1967). *Nature, Lond.* **216**, 567.
Pacini, F. (1968). *Nature, Lond.* **219**, 145.
Richards, D. W. & Comella, J. M. (1969). *Nature, Lond.* **222**, 551.
Staelin, D. H. & Reifenstein, E. C. (1968). *Science* **162**, 1481.
Willstrop, R. V. (1969). *Nature, Lond.* **221**, 1023.

2
Techniques for search and for observation

2.1 The problem of sensitivity

Catalogues of discrete radio sources, obtained from surveys of several steradians of the sky, usually have a lower limit of sensitivity of the order of 1 Jy (1 Jy = 1 Jansky = 10^{-26} W m^{-2} Hz^{-1}). Surveys reaching lower flux densities achieve an improved sensitivity only by the use of long integration times, which in the most sensitive aperture synthesis techniques extend to many hours of observation. By coincidence, the mean flux density at metre wavelengths of the strongest pulsars is approximately 1 Jy. The failure of the radio source surveys (apart from Hewish's scintillation survey) to detect any of the pulsars is therefore not surprising; it is, however, very remarkable that search techniques for pulsars now commonly extend to flux densities of the order of 0.01 Jy, and that once a pulsar has been found its characteristics may be measured in some considerable detail despite the weakness of its signals.

The background electrical noise level in a typical large radio telescope corresponds to a flux density of order 100 Jy: this might for example be obtained in a telescope of the size of the 250-foot Mk I radio telescope at Jodrell Bank, operating with a system noise temperature of 100 K. The mean sensitivity required for pulsar observations is therefore 10^{-4} of the noise level. A typical pulsar might, however, only radiate strongly for about 2% of its period, so the peak pulse signal is fifty times greater than the mean. The flux density during the pulse would then be $\frac{1}{2}$ Jy. The sensitivity of the receiver is determined by its bandwidth B and output integration time τ: typically a receiver might have $B = 1$ MHz and $\tau = 10$ ms, giving a sensitivity $(B\tau)^{1/2} = 10^2$ better than the input noise level, so that the system would detect a single pulse with flux density considerably larger than 1 Jy lasting more than 10 ms. We see that even if the integration time of the receiver exactly matched the pulse duration this system would not detect individual pulses from the typical pulsar. The detection of detail within the pulses would require an even shorter integration time, which worsens the sensitivity.

Evidently only the largest radio telescopes can be used for pulsar searches. Furthermore, systems which rely on the detection of individual pulses, as was the case in the initial discoveries and for the Crab Pulsar, have very limited sensitivity. We shall see that some improvement can be obtained by increasing the radio bandwidth, despite a limitation caused by interstellar dispersion, but the main gain is obtained in practice by increasing the integration time. This increase is achieved by the superposition of many pulses, using the characteristic periodicity of pulsar signals as a means of signal recognition.

The techniques of pulsar searches and observations depend heavily on signal processing, which is essentially the art of extracting weak signals with known distinguishing characteristics out of a background of random noise. The two important characteristics are periodicity and frequency dispersion in arrival time, both of which have been used in search techniques.

2.2 Frequency dispersion in pulse arrival time

Radio pulses travel in the interstellar medium at the group velocity v_g. In an ionised gas the group velocity is related to the phase velocity v_p by

$$v_g v_p = c^2.$$

The phase velocity at wavelength λ is obtained from the refractive index μ_p:

$$\mu_p = \left(1 + \frac{Nr_0\lambda^2}{\pi}\right)^{1/2}$$

where $r_0 = e^2/mc^2$ is the classical radius of the electron, and N is the electron density. Hence for small N

$$v_g = \left(1 - \frac{Nr_0\lambda^2}{2\pi}\right)c$$

and the delay t in travel time over a distance L, compared with free space, is

$$t = \frac{Nr_0 c \nu^{-2}}{2\pi} L.$$

$$= 1.345 \times 10^{-3} \nu^{-2} NL \text{ s Hz}^{-2} \text{ cm}^{-2}.$$

It is convenient to characterise the transmission path by the product NL, which in customary astrophysical practice is measured in units pc cm^{-3}. The product is then known as dispersion measure DM. Observers commonly quote radio frequencies in megahertz, so that the delay becomes

$$t = 4.15 DM \nu_{MHz}^{-2} \text{ s}.$$

2.3 Search techniques: single pulses

The frequency dependence of this delay has a very important effect on observations of radio pulses. A short broad-band pulse will arrive first at high frequencies, traversing the spectrum at rate

$$\dot{\nu} = -\frac{\nu_{MHz}^3}{8.3 \times 10^3 DM} \text{ MHz s}^{-1}.$$

Correspondingly, a pulse with length τ occupies instantaneously a bandwidth

$$B_i = \frac{\nu_{MHz}^3}{8.3 \times 10^3 DM} \tau \text{ MHz}.$$

A receiver with a smaller bandwidth than B_i will not receive the whole of the pulse energy, and the sensitivity will be reduced. If the receiver has a larger bandwidth B_r the pulse will be detected for an extended time, while the frequency dispersion takes the pulse energy across the receiver band. The effect is that the receiver sensitivity is proportional to $(B_r^{-2} + B_i^{-2})^{1/4}$. This factor was applied by Large & Vaughan (1971) in the calibration of the Molonglo survey, in which the detection of pulsars depended on the recognition of single pulses.

The effect of dispersion is to reduce the sensitivity of a search with fixed receiver bandwidth B_r to pulsars with high dispersion. The lost sensitivity may, however, be recovered by dividing the receiver bandwidth B_r into separate bands each at least as small as B_i, and using separate receivers on each band. The output of these separate receivers can then be added with appropriate delays so that the pulse components are superposed. This process of 'de-dispersion' is illustrated in Fig. 2.1.

Various techniques have been used to achieve 'de-dispersion'. Large & Vaughan (1971) used a mechanically driven sequential sampling of the separate receiver outputs: more usually a digital delay system has been used by several observatories. The difficulty in a search process is, of course, that the system must be 'tuned' for an arbitrary range of dispersion measure DM, which will necessarily select a certain class of pulsars. Digital delay methods may, however, be set to a sequence of delays, operating on a single set of input signals, provided that sufficient computer capacity is available.

2.3 Search techniques: single pulses

All the pulsars discovered in the Cambridge survey, and all those discovered in the southern hemisphere survey at Molonglo, were detected by the appearance of single pulses on pen chart recordings. The Molonglo radio telescope proved to be particularly well suited for pulsar

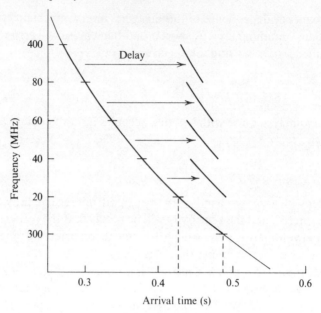

Fig. 2.1. De-dispersion. Arrival time versus frequency for a pulsar with dispersion measure $DM = 100$ pc cm^{-3}. A sharp pulse is lengthened to 0.21 s in a receiver covering 300 to 400 MHz. If separate receivers are used for each 10-MHz band, with each output delayed as shown, the pulse can be received within 0.06 s, giving an improvement in sensitivity.

search, and twenty-eight pulsars were found by inspection of survey recordings. Simple pulse-lengthening techniques were used, to compensate for the slow response time of the normal chart recorders (Vaughan & Large, 1969). The survey was extended to higher values of dispersion measure DM by the use of a de-dispersion technique involving twenty receiver channels; only two pulsars were found by this technique, presumably because the larger values of DM corresponded to the more distant pulsars, whose detection is limited by sensitivity alone.

In some searches the dispersion effect has itself been used as a means of recognition of individual pulses against a background of impulsive electrical noise and interference. The discovery of the Crab Pulsar by Staelin & Reifenstein was made in a survey using fifty receiver channels covering 110 MHz to 115 MHz. The Crab Pulsar is unusual in its production of occasional very strong individual pulses (the 'Giant' pulses: see Chapter 11). Some of these showed on the recordings, with a dispersion delay of over a second in the whole bandwidth. The dispersion of the Crab Pulsar ($DM = 57$) is in fact so large that even within a single channel 100 kHz wide a regular pulsation would be smeared out over the 33-ms period; the discovery depended entirely on the individual giant pulses and not on the periodicity.

A more extensive search for dispersed pulses made at Jodrell Bank used two receiver channels separated by 2 MHz at 408 MHz. Three new pulsars were discovered in this search (Davies & Large, 1970).

2.4 Search techniques: periodicity

The most successful searches have used the precise periodicity of pulsars as their distinguishing characteristic. Searches using this principle have detected all the known pulsars within a given search area, as well as finding new ones; nevertheless it must be remembered that some pulsars are very erratic in their emission, so that their pulses would not be as easily detected in a 'periodicity search' as would the completely regular pulses from most other pulsars. This effect may possibly be operating as a selection effect in favour of pulsars with regular pulse characteristics.

The principle of a periodicity search may be either to find a train of regularly spaced pulses, or to look for a sinusoidal Fourier component. The two approaches are closely related: in fact the pulse train is best regarded as a sum of Fourier components, fundamental plus harmonics. Evidently there is more information in the pulse train, giving a higher signal-to-noise ratio than the fundamental alone. However, surveys are usually made with large receiver input bandwidths, so chosen that a large dispersion measure will smear out the pulses into an approximate sinusoid without high harmonics. Since the Fourier analysis is also more economical of computer time, it is usually judged not to be worthwhile to include the harmonics and conduct a full pulse search.

The computer capacity required for a pulsar search has been studied by Lovelace, Sutton & Salpeter (1969). The noise-like signal at the receiver output is recorded as a series of N samples at interval Δt, over a total time $N\Delta t$. The sampling interval must be smaller than half the shortest period to be covered in the search, so that the available range of periods ranges from $2\Delta t$ to $N\Delta t$. These samples must then be searched for periodic pulses, or for the fundamental Fourier component of the pulse train. This is achieved by a correlation analysis. For example, to test for the presence of a pulse Δt wide, with period $p\Delta t$, one should multiply the data samples by an idealised pulse train, repeating the process in p different phases, and searching for the largest outputs of this correlation process. This must then be repeated for the whole available range of p. In practice this is very extravagant of computer time, even when economical algorithms have been devised to minimise the number of steps in the computation. The

Techniques for search and observation

usual practice is to correlate the pulse train only with a fundamental cosine and sine wave.

The sensitivity of a search is determined by the total length of the data $N\Delta t$, which becomes the integration time. The search is therefore conducted by increasing N until the computation time for N samples becomes approximately equal to the data length. The number of distinguishable periods in a data string with N points is $N/2$. If each period were to be searched for separately, each would require N summations, so that a total of $N^2/2$ would be required for the search. In fact the search may be conducted by a series of folding operations, so that the basic summations total $(N/2) \log N$.

The periodicity search at Jodrell Bank (Davies, Lyne & Seiradakis, 1976) used data trains 10 min long, containing 12 000 samples each representing 40 ms of signal. During the recording of data, the previous data train was being analysed, and largest Fourier components were being printed out. Using the Mk IA 250-foot radio telescope, pulsars were discovered by this method at the rate of about one per 2 days of observation. The total discovered amounted to thirty; another twenty-one pulsars previously known also appeared in the search, so that a considerable sample was available from the one search technique for statistical analysis of pulsar populations (Chapter 20).

A good impression of the difficulty of detecting many more pulsars is obtained by considering the technical problem of finding a pulsar in M31, the Andromeda Nebula, a project which would have great astrophysical interest. The present level of sensitivity is such that the most powerful pulsar in our Galaxy can be detected in the Jodrell Bank search procedure if it is at about the distance of the Galactic centre. If it were placed in M31, the signal would be reduced by about 35 decibels, and the signal-to-noise ratio at the receiver could only be restored by increasing the integration time by a factor of over 10^7, i.e. to over 3 years. The computation to search for precise periodicities must then allow for annual variations due to the Earth's motion, but a full search through the whole range of periodicities would in any case take a considerably longer time than the observation itself.

2.5 Optical, X-ray and gamma-ray techniques

At optical wavelengths, and for all shorter wavelengths, pulsar recording techniques involve photon counting, and the sensitivity and accuracy of any measurement depend entirely on photon statistics. A special class of pulsating source has been discovered among the discrete X-ray sources; these X-ray pulsars differ in many ways from the other pulsars (see

2.5 Optical, X-ray and gamma-ray techniques

Chapter 4), and we shall here confine our attention to the only example of a pulsar which radiates throughout the whole available electromagnetic spectrum, i.e. the Crab Pulsar.

The mean level of visible light from the Crab Pulsar corresponds to visible magnitude $V = 16.5$ (Kristian, 1971). The aperture of a 1-metre telescope should therefore collect about 200 photons per pulse in visible light; with the usual light losses and inefficiency of detection there may be only about ten photons actually detected per pulse. A detailed light curve may involve a time resolution of about 50 μs; in this time interval the probability of receiving a photon is much less than unity even at the peak of the pulse. The light curve can in fact only be constructed by the superposition of a large number of pulses, typically over 10^4 pulses (i.e. more than 5 minutes' integration).

X-ray telescopes typically have a smaller collecting area, of order 0.1 m^2; furthermore the energy of a photon is about 10^3 greater than in visible light, so the sensitivity of the system is typically 10^4 lower than that of the 1-metre telescope. The detective quantum efficiency of the system, however, almost reaches unity, and the spectral bandwidth is large. There is over 100 times more power radiated by the pulsar in the X-ray spectrum than in visible light. The result is that about one or two photons are detected per pulse in a typical rocket-borne X-ray telescope. With such a small counting rate, it is fairly difficult to draw out a precise 'light curve' in the limited duration of an X-ray rocket flight. Satellite-borne telescopes will, of course, improve on this situation.

Gamma-ray astronomy covers a large range of the spectrum, corresponding to energies from 500 keV to 5×10^{13} eV. Techniques at the lower energies (up to 10 MeV) involve spark chambers, nuclear emulsions, or Cerenkov detectors, mounted in balloons or satellites. There is good evidence that the Crab Pulsar has been detected in this energy range (Hillier *et al.*, 1970). At much higher energies a different technique exists for ground-based observations. This depends on the detection of Cerenkov light pulses emitted by extensive air showers which occur when gamma-rays with energy greater than about 10^{13} V fall on the atmosphere (Charman, Jelley & Drever, 1970). Wide-angle light detectors can be used, so that the effective collecting area of this atmospheric detector system may be several square kilometres; furthermore, the detective quantum efficiency may be large, since practically every gamma-ray should give rise to an extensive air shower. The most exhaustive search using this technique has been carried out at Mount Hopkins (Fazio *et al.*, 1971). Only upper limits to the flux of gamma-rays from the Crab Pulsar have as yet been established.

Techniques for search and observation

The extension of detailed measurements of the integrated pulse profile, and of the polarisation of pulses from the Crab Pulsar, into the X-ray and gamma-ray region is difficult but very important in the theory of pulse formation. There is considerable scope for new techniques to cover more of the 40 octaves spanned by the electromagnetic radiation from this remarkable object.

References

Charman, W. N., Jelley, J. V. & Drever, R. W. P. (1970). *Acta Phys., Acad. Sci. Hung. Suppl.* **1**, 63.

Davies, J. G. & Large, M. I. (1970). *Mon. Not. R. astron. Soc.* **149**, 301.

Davies, J. G., Lyne, A. G. & Seiradakis, J. (1976). *Mon. Not. R. astron. Soc.* (in press).

Fazio, G., Hearn, D., Helmken, H., Rieke, G., Weekes, T. & Chaffe, F. (1971). *IAU Symposium No. 46*, p. 160. (Dordrecht: D. Reidel.)

Hillier, R. R., Jackson, W. R., Murray, A., Redfern, R. M. & Sale, R. G. (1970). *Astrophys. J.* **162**, L177

Kristian, J. (1971). *IAU Symposium No. 46*, p. 87. (Dordrecht: D. Reidel.)

Large, M. I. & Vaughan, A. E. (1971). *Mon. Not. R. astron. Soc.* **151**, 277.

Lovelace, R. V. E., Sutton, J. M. & Salpeter, E. E. (1969). *Nature, Lond.* **222**, 231.

Vaughan, A. E. & Large, M. I. (1969). *Proc. Astron. Soc. Australia* **1**, 220.

3
The identification with rotating neutron stars

Although the grounds for the identification of pulsars with neutron stars are both secure and straightforward, it is of considerable interest to trace the steps which led to this identification in more detail than is contained in the brief history of Chapter 1. The interest is twofold. First, the story forms an introduction to the major topics in pulsar astrophysics with which this book is concerned. Second, there were among the early theories some interesting and basic physical ideas which, although rejected at the time, may re-emerge with different applications. In retrospect, such ideas as oscillating white dwarf stars and rapidly orbiting binaries, seem wide of the mark; nevertheless, the discovery of the pulsars did stimulate new work on oscillations, involving a re-examination of the equation of state of condensed matter, while the binary theory soon found practical application in the X-ray pulsars and, later, in the relativistic dynamics of the binary pulsar discovered in 1974.

3.1 Oscillations of condensed stars

Precisely periodic behaviour is expected in astronomy from rotation or orbital motion in dynamical systems, and from stellar oscillations. The dynamical theories were not easy to accept originally, because the periods were so short that white dwarf stars could be ruled out, leaving only the unfamiliar neutron stars. There was less difficulty in accepting the idea of oscillation, since there had recently been a study of the radial oscillation of condensed stars which gave periodicities not far different from those of the first few pulsars to be discovered.

In 1966 Melzer & Thorne showed that a white dwarf star could have a resonant periodicity of about 10 s, for radial oscillations in the fundamental mode. This would correspond to the oscillations actually observed in the Cepheid variable stars, except that the time scale was shorter by three orders of magnitude. In the Cepheids the oscillation is driven by the energy of nuclear processes at the centre of the star; the driving process depends on the transfer of energy to pressure and temperature variations through a cyclic change in the opacity of the stellar material.

The period is determined mainly by the resonance of the star, which is determined partly by gravity and partly by the compressibility, i.e. by the equation of state of the material. Melzer & Thorne found that the period for a white dwarf depends on the central density in such a way that a minimum period was obtained in the centre of the allowable mass range, where the central density is about 10^7 g cm^{-3}. Their value for the minimum period was 8 s.

The period is easily calculated if the oscillation is controlled entirely by gravity. Dimensional arguments show that the period is independent of radius, and proportional to $(G\rho)^{-1/2}$ where ρ is the density. If compressibility or elasticity becomes important, then the velocity of sound waves increases and the period is reduced. Elasticity is the dominant factor for small solid bodies, whose diameters are not determined by their own self-gravitation. Densities of order 10^7 g cm^{-3} correspond to periods of order 10 s for purely gravitational oscillations in white dwarf stars, so that no very large contribution from elasticity seemed to be necessary to explain pulsar periods of about 1 s. The neutron star material behaves somewhat differently from a gas under adiabatic oscillations; furthermore, some correction must be made for relativistic effects. The theoretical period was soon reduced, but only down to 2 s.

A further reduction in the theoretical period of white dwarf oscillations was only possible by changing the geometry. The first such proposal was obviously to involve harmonic modes. The overtone modes of oscillation are not in precise harmonic ratio, but a low-order harmonic could easily reach a period below 1 s. The proposal was, however, artificial, since there was no accompanying suggestion for exciting overtones without the fundamental (it had also to be admitted that there was no good theory even for exciting the fundamental).

A second geometric proposal was to change the shape of the star, by considering the distortion due to rapid rotation. A highly flattened rotator would oscillate in more complex modes, but a simple fundamental mode corresponding to radial oscillation along the minor axis would have a shorter period than for the spherical star. This proposal (Ostriker & Tassoul, 1968) reached down to periods below 1 s.

The difficulties encountered in reducing the theoretical oscillation periods of white dwarf stars down to 1 s and below were matched by the difficulty in accounting for these periods as oscillations of neutron stars. Melzer & Thorne had included the calculations in their classic paper. The fundamental modes of radial oscillation had periods in the range 1 to 10 ms, and no possibility seemed to exist for lengthening the periods by two orders of magnitude.

3.2 Planetary and binary orbits

At this point in the argument the oscillation theories were overtaken by the discoveries of the short-period pulsars. The periods of the Vela Pulsar (88 ms) and of the Crab Pulsar (33 ms) lay in the middle of the impossible gap between white dwarfs and neutron stars, and the attempts to stretch the theoretical periods to cover the observed periods of the pulsars were abandoned.

3.2 Planetary and binary orbits

Let us suppose that the pulsar period T is the orbital period of a planet, or satellite, in a circular orbit, radius R, round a condensed star with mass $M_\odot$ (in solar units). Then

$$R = 1.5 \times 10^3 M_\odot^{1/3} T^{2/3} \text{ km}. \tag{3.1}$$

It is therefore just possible for a satellite to orbit a white dwarf star with a period of 1 s, but the orbit would be grazing the surface. It would be more reasonable to consider a neutron star as the central object, when periods down to 1 ms would be possible. There are, however, two insuperable objections to the proposition that orbiting systems of this kind provide a model for pulsars.

The main difficulty concerns gravitational radiation, which is due to the varying quadrupole moment of any binary system. The energy loss through gravitational radiation would lead to a decrease in orbital period. A general formulation of the time scale τ of this change was given by Ostriker (1968) for a binary system with masses M and εM, with angular velocity ω:

$$\frac{1}{\tau} = \frac{1}{\omega}\frac{d\omega}{dt} = \frac{96}{5}\frac{\varepsilon}{(1+\varepsilon)^{1/3}}\frac{(GM)^{5/3}}{c^5}\omega^{8/3}. \tag{3.2}$$

For a satellite mass m where $\varepsilon = m/M$ is small, and $M = M_\odot$,

$$\tau = 2.7 \times 10^5 \varepsilon \text{ s}. \tag{3.3}$$

The time scale was evidently far too short unless the satellite mass was very small. Pacini & Salpeter (1968) soon established that early observations of the stability of the period showed that m must be less than 3×10^{-8} solar masses.

Even the improbable hypothesis that such a small mass could be responsible for the radio pulses faced a second problem. The satellite would be orbiting in a very strong gravitational field, which would tend to disrupt it by tidal forces. Pacini & Salpeter showed that even if it were made of high tensile steel it could not withstand these forces unless it was smaller than about 20 m in diameter. An added problem would be that the satellite would be liable to melt or evaporate in the very high radiation field of a pulsar.

The same situation evidently obtained *a fortiori* for a binary system, for which a very rapid change in period would be expected. Planetary and binary systems were therefore eliminated as possible origins for the clock mechanism of pulsars. Gravitational radiation itself does, however, recur in the pulsar story; a pulsar was eventually found which was itself a member of a binary system with the short period of $7\frac{3}{4}$ hours, in which the orbital period should decrease due to gravitational radiation at the rate of 30 ms per year (see Chapter 7). If this can be detected (which admittedly is doubtful) it will provide a much-needed test of the theory of gravitational radiation.

3.3 Rotating condensed stars

The maximum angular velocity ω of a spinning star is determined by the centrifugal force on a mass at the equator. An estimate is easily obtained by assuming that the star is spherical with radius r; the centrifugal force is then balanced by gravity when

$$\omega^2 r = \frac{GM}{r^2}. \tag{3.4}$$

This is, of course, the same condition as for a satellite orbit grazing the surface. If the star has uniform density ρ, the shortest possible rotational period P_{min} is

$$P_{min} = 3^{1/2} \pi^{1/2} (G\rho)^{-1/2}. \tag{3.5}$$

A period of 1 s therefore requires the density to be greater than 10^8 g cm^{-3}, which is just within the density range of white dwarf stars. Neutron stars, on the other hand, can rotate with a period as small as 1 ms.

The limit on rotational angular velocity is somewhat more severe than in this simple argument, because the star will distort into an ellipsoid and tend to lose material in a disc-like extension of the equatorial region. The white dwarf theory was therefore already on the verge of impossibility for the first pulsars; the discovery of the short-period pulsars at once ruled it out completely.

The probable identification with rotating neutron stars then led to the interpretation of the radio pulses as a 'light-house' effect, in which a beam of radiation is swept across the observer. This idea was supported by the observation that the plane of polarisation of radio waves from the Vela Pulsar swept rapidly in position angle during the pulse, which agreed with

3.4 Changes in period

some simple models of beamed emission. The radio source must then be localised, and directional, as well as powerful. This led Gold (1968) to his seminal note in *Nature*, in which he suggested the identification with rotating neutron stars, the existence of a strong magnetic field which drove a co-rotating magnetosphere, and the location of the radio source within the magnetosphere, probably close to the velocity-of-light cylinder. He also pointed out that rotational energy must be lost through magnetic dipole radiation, so that the rotation would be slowing down appreciably.

3.4 Changes in period

The early measurements of period on the first pulsar CP 1919 showed that no change was occurring larger than 1 part in 10^7 per year. This limit was very close to the actual changes which were measured a few years later, but the early null result could be used only to show that the stability of the period was in accord with the large angular momentum of a massive body in rapid rotation. Pacini (1968) showed that the limit on slowdown implied a magnetic field strength at the poles of a white dwarf less than 10^{12} gauss (10^8 tesla). He considered only magnetic dipole radiation in free space, which radiates away the rotational energy at a rate

$$\frac{dW}{dt} = -\frac{2\pi\omega^4\mu_0}{3c^3}r^6 H_0^2 \sin^2\alpha \qquad (3.6)$$

where H_0 is the polar field and α is the angle between the dipole axis and the rotation axis.

The slowdown of the Crab Pulsar was first measured by Richards & Comella (1969). From October 1968 to February 1969 the period lengthened uniformly by 36.48 ± 0.04 ns per day, i.e. by over 1 μs per month. The rate of change was consistent with the known age of the Crab Nebula, confirming the association of the pulsar with the supernova explosion. Furthermore, the rate of change could be applied to the neutron star theory, giving an energy output which was sufficient for the excitation of the continuing synchrotron radiation from the Crab Nebula. This coincidence was the final proof of the identification, as pointed out in Gold's second *Nature* letter (1969).

In retrospect, it is intriguing to consider what deduction might have been made from the measured variations of rotation period of the Vela Pulsar, if it had happened (as it nearly did) that those measurements had preceded those of the Crab Pulsar. The period of the Vela Pulsar was observed to be increasing slowly from November 1968 to February 1969, at the rate of 11 ns per day, but at the end of February a discontinuous

Identification with rotating neutron stars

decrease in period occurred, amounting to 200 ns. The change occurred in less than a week (Radhakrishnan & Manchester, 1969; Reichley & Downes, 1969). By the time that this anomalous step was announced, the neutron star theory was already firmly established, and the decrease in period was regarded as an aberration rather than the typical behaviour, as might be expected in orbiting systems. The step, or 'glitch', was interpreted solely on the basis of a change of moment of inertia, due to shrinkage or a 'starquake' (Chapter 7).

References

Gold, T. (1968). *Nature, Lond.* **218**, 731.
Gold, T. (1969). *Nature, Lond.* **221**, 25.
Melzer, D. W. & Thorne, K. S. (1966). *Astrophys. J.* **145**, 514.
Ostriker, J. P. (1968). *Nature, Lond.* **217**, 1127.
Ostriker, J. P. & Tassoul, J.-L. (1968). *Nature, Lond.* **219**, 577.
Pacini, F. (1968). *Nature, Lond.* **219**, 145.
Pacini, F. & Salpeter, E. E. (1968). *Nature, Lond.* **218**, 733.
Radhakrishnan, V. & Manchester, R. N. (1969). *Nature, Lond.* **222**, 228.
Reichley, P. E. & Downs, G. S. (1969). *Nature, Lond.* **222**, 229.
Richards, D. W. & Comella, J. M. (1969). *Nature, Lond.* **222**, 551.

4
The X-ray pulsars

There are great practical difficulties in making astronomical observations in the X-ray region of the electromagnetic spectrum. Atmospheric absorption makes it necessary to observe from a rocket or satellite, with all the concomitant limitations of apparatus weight and size, and the need to prepare any large-scale experiment with great care, often many years in advance of launching. The counterparts of familiar optical instruments such as reflecting telescopes and spectrometers are hard to construct and limited in performance. Until recently, there was the further discouragement that the X-ray fluxes from known visible objects – the planets, stars and the visible surface of the Sun – would be too weak to be detectable.

The first X-ray astronomical observations were of the Sun, for there was good evidence accumulating that there should be observable X-radiation from disturbed regions of the chromosphere and corona. Then a flux of X-rays from the Crab Nebula was discovered, as expected from the synchrotron theory which had already been developed to explain the visible and radio emission. The most interesting sources of X-rays were, however, almost completely unexpected. These are the stellar sources, which include the X-ray pulsars.

For detectable X-rays to be emitted from thermal electrons in the atmosphere of a star, it is necessary for a dense part of the atmosphere to be heated to a temperature of at least 10^6 K. In the Sun only the tenuous outer corona is at such a high temperature, and X-rays are only seen when the lower part of the atmosphere is heated, as by a solar flare, or when a disturbance increases the density in the corona. A prediction of a far more powerful thermal source was made in 1964 independently by Zel'dovich & Guseynov and by Hayakawa & Matsouka. These authors introduced the concept of binary star systems as X-ray sources.

If in a binary star system one star is a condensed object, such as a white dwarf or a neutron star, and the other is a more massive normal star which is losing mass through a stellar wind, there might be a very large rate of accretion onto the condensed star, and a hot, dense atmosphere would then develop. Plenty of energy would be available to heat the accreted

The X-ray pulsars

particles: the gravitational energy corresponds to several kilovolts for a white dwarf, and about 100 MeV for a neutron star.

The first X-ray source which was definitely identified with a condensed object was the Crab Pulsar. This pulsar is not, however, a member of a binary system, and the pulsed emission is not thermal (Chapters 11 and 16). The certainty of the identification rested on the discovery of the periodic pulsed emission. There had been earlier discoveries which seemed to fulfil the predictions of the binary model, but the evidence was insufficient. Notably the strongest X-ray source in the sky, Sco X-1, was found to be variable on short time scales; further, it was identified with a visible star whose spectrum fitted that expected from the binary partner undergoing mass loss. Sco X-1 was also found to be a radio source.

Despite the discovery of several other X-ray sources in a series of rocket observations, there was for several years no further evidence available for the binary nature of Sco X-1 or any other source. The situation changed dramatically with the launch of the satellite Uhuru (the Swahili word for freedom, commemorating its launch from the coast of Kenya). This X-ray satellite gave a catalogue of 161 sources, including two which are pulsating with short precise periodicities. These are the two prototype X-ray pulsars, Hercules X-1 and Centaurus X-3. The Crab Pulsar is now regarded as an exception, outside the category of X-ray pulsars.

The Uhuru catalogue contains several different classes of source. Fig. 4.1 shows the distribution in observed strength and in galactic latitude. The strong sources at low latitudes are within the Galaxy, while the weaker ones at high latitudes are partly, and perhaps wholly, extragalactic. Some strong extragalactic sources are identifiable with abnormal galaxies such as the Seyfert galaxy NGC 1275; other weaker ones emit as though they contained the same collection of discrete sources as are found in our own Galaxy. Among the galactic sources are several of the known supernova remnants. This leaves about forty other galactic sources, all of which are now believed to be associated with binary systems. Although there probably is a wide variety within this class, it is instructive first to look in detail at the two clear X-ray pulsars, and then to see if the other sources may be placed in the same category.

4.1 Her X-1/HZ Her

Her X-1 is a powerful X-ray source which is known to be a member of a binary system, the other star being identified with a bright visible star HZ Her. Membership of the binary system was proved by the observation of the periodic occultation of the X-ray source, at intervals of 1.7 days

4.1 Her X-1/HZ Her

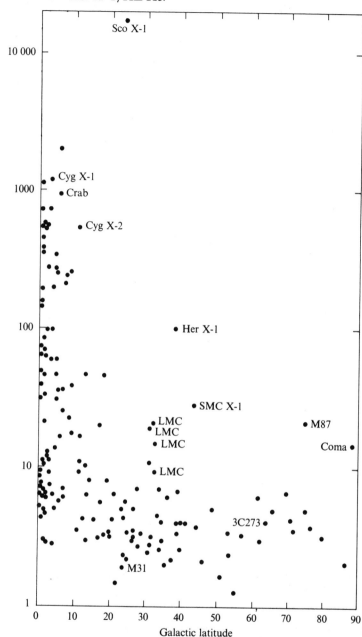

Fig. 4.1. The Uhuru X-ray sources. The strongest sources are concentrated close to the plane of the Galaxy. These include such galactic sources as Cyg X-1, the Crab Nebula, and Sco X-1. Extragalactic sources such as the Magellanic Clouds (LMC and SMC), M87, and the Coma Cluster are found at higher galactic latitudes. (After Giacconi, 1974.)

The X-ray pulsars

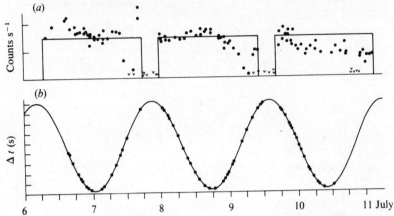

Fig. 4.2. The X-ray source Her X-1. (*a*) Periodic occultation, (*b*) Sinusoidal variation of arrival time. (After Giacconi, 1974.)

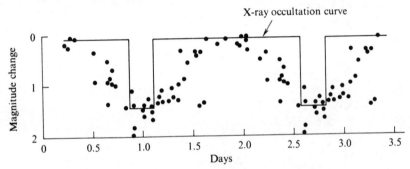

Fig. 4.3. Periodic optical variation of the star HZ Her, superposed on the X-ray occultation curve of Her X-1. The optical magnitudes cover years between 1944 and 1972. (After Giacconi, 1974.)

(Fig. 4.2). The X-radiation is completely cut off for 0.24 days, with sharp transitions between the on and off states. Many photographic plates of the star HZ Her exist in the files, dating back over 50 years, but it was only in response to the X-ray discovery that an optical variability with the same period of 1.7 days was found. The comparison of the X-ray and visible cycles in Fig. 4.3 establishes conclusively that the two objects are binary partners.

Her X-1 also has another regular feature. The X-radiation is pulsed at a period of 1.24 s, as seen in Fig. 4.4. This short period leads, through the arguments of Chapter 3, directly to its identification with a neutron star. The precise clock mechanism giving the period of 1.24 s is the rotation of the star. A remarkable and beautiful observation was now made which linked the 1.7-day and 1.24-s periods. The 'pulsar clock' was supposed to

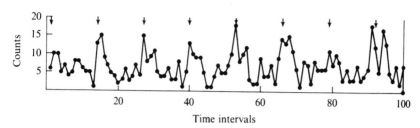

Fig. 4.4. X-ray pulse counts from Her X-1. Each count is in an interval of 0.096 s. The arrows indicate the periodic peaks at 1.24-s intervals. The small counts also show random statistical fluctuations. (Uhuru data.)

be in orbit round a heavy star. There should then be a variation in distance from the observer, so that the pulses arrive early when the neutron star is closest, and late when it is most distant. Between these times the clock would appear to run slow as the star receded and went into eclipse, and fast as the star came out of eclipse travelling towards the observer. This effect is observed as a sinusoidal variation of arrival time, fitting exactly in phase with the occultation.

The good fit to a sine wave shows that the orbit is nearly circular (eccentricity <0.05). The delay time is 26.4 s (peak to peak), so that the radius of the orbit is $4.10^6 \sin i$ km where i is the inclination of the pole of the orbit to the line of sight.

A third regular feature of the X-radiation only appears after a longer period of observations. The strength of the signal varies with a period of 35 days, so that it is visible for about 11 days and invisible for 24 days. This is not so easily explained in terms of the dynamics of the binary system. It may be due to free precession of the neutron star, but it might alternatively be due to a periodic effect on the companion star. The latter seems unlikely since there is no corresponding periodicity in the optical luminosity. If it is precession, there are interesting consequences for the interior structure of the star, since there must be an appreciable ellipticity. Precession, however, must explain not only the 35-day period; it must provide an explanation for the modulation of the X-ray intensity.

4.2 Cen X-3

The observational evidence for the binary nature of this source rests solely on the X-ray timing observations. As for Her X-1, there is a short pulse period, which shows a sinusoidal variation of pulse period with an occultation cycle (Fig. 4.5). The pulse period is 4.8 s, and the binary period is 2.087 days. The amplitude of the period sinusoid is 0.0067 s.

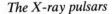

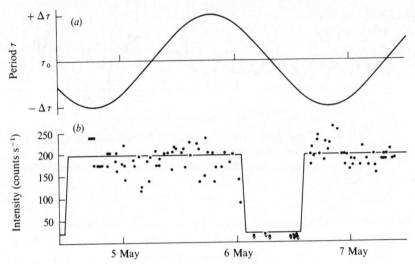

Fig. 4.5. The binary X-ray source Cen X-3. (*a*) Variation of pulse delay. (*b*) Occultation of X-rays. (After Giacconi, 1973.)

This was the first X-ray source observed to be pulsating (Giacconi *et al.*, 1971). It does not have a known visible counterpart, and it does not have a long-term cyclic variation like the 35-day cycle of Her X-1. Nevertheless, it does seem to be very similar to Her X-1, and we may assume that both sources are indeed close binaries, with a massive primary losing mass to a neutron star secondary.

4.3 Period changes and mass transfer

All the radio pulsars derive their energy from their kinetic energy of rotation, and their periods are all steadily increasing as the rotation rate decreases. Her X-1 and Cen X-3 show the opposite effect, with an irregular but comparatively rapid decrease in period. Fig. 4.6 shows the decrease over the first 2 years of observation.

The existence of a decrease immediately confirms the postulate of mass transfer in the binary system. The magnitude of the effect depends on the details of the accretion: the theory requires a knowledge of the distance from which particles carrying angular momentum are brought in to the star. But the rate nevertheless gives a time scale, or lifetime for the phenomenon. Cen X-3 shows a very large decrease, amounting to 3 ms in 2 years. At this rate the lifetime is only about 10^3 years. Her X-1 has a smaller decrease, amounting to 6 μs per year on average. The lifetime is then about 10^5 years. These short lifetimes are easily reconcilable with the theory of stellar evolution in binary systems (Chapter 19).

4.3 Period changes and mass transfer

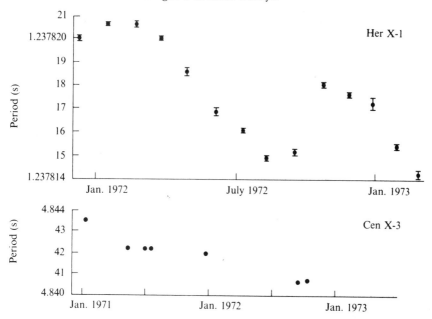

Fig. 4.6. Changing periods of X-ray pulsations from Her X-1 and Cen X-3. (After Giacconi, 1974.)

A further effect of mass transfer is to change the orbital period. The way in which this changes is also affected by the details of mass loss from the normal star. At one extreme there may be a stellar wind type of outflow with very little material ending up on the neutron star, while at the other extreme there may be a steady flow confined to the region of the gravitational neutral point between the stars, when all the lost material might end up on the neutron star. The material would probably not fall directly on to the surface, but it would form a disc from which material would fall to the surface. Without the formation of this accumulation disc, i.e. if the matter is directly accreted, the period P would change at a rate $\dot{P}$ given by

$$\dot{P} = 3P\left(\frac{M_1 - M_2}{M_1 M_2}\right)\frac{dM_1}{dt} \qquad (4.1)$$

where M_1 is the mass of the large star and M_2 that of the neutron star. If the matter is accreted on to a disc at radius r instead of at the surface then $\dot{P}$ is reduced by the factor $(r/r_0)^2$. Other models have been worked out, e.g. by van den Heuvel & de Loore (1973).

The behaviour of the rotation periods of Cen X-3 and Her X-1 are shown in Fig. 4.6. Although there are large changes in rotation period,

The X-ray pulsars

the reversal in Her X-1 makes it impossible to fit a single model. All that can be said is that the time scale, and the order of magnitude of the changes, are both consistent with the expected mass loss rates of evolving heavy stars, i.e. with rates of 10^{-5} or 10^{-6} solar masses per year.

4.4 Other known X-ray binaries

No other pulsating sources are known, so that no other orbital information is available via X-rays. But there are several other optical identifications, which fairly certainly represent the binary counterparts of neutron stars. These are included in the table of binary X-ray sources (Table 4.1). Cyg X-1 has an orbital period derived originally from spectroscopic variability in the optical star, and found later as a variation in strength of the soft X-rays (but not of the hard X-rays). It shows quasi-periodic pulsations on time scales as short as 50 ms, but has no clearly defined period. It is a particularly interesting object: calculations of the mass of the condensed component suggest that it may be larger than the upper limit of $1.7 M_\odot$ allowable for a neutron star (Chapter 5). The inevitable conclusion would then be that it is in the next, and final, state of collapse, namely a 'black hole'.

TABLE 4.1 *Binary X-Ray sources*

Source X-ray/optical	Pulsation period (s)	Binary period (d)	Other time structure	Optical star type	Peculiarities
Her X-1/HZ Her	1.2378	1.700	35-day period	A9 or F0	High galactic latitude. Late type star
Cen X-3/—	4.842	2.087	–	–	Rapid period decrease
Cyg X-1/HDE 226868	–	5.600	Rapid X-ray fluctuations	B0 super-giant	–
3U 1700-37/ HD 153919	–	3.4	Rapid X-ray fluctuations	07	–
Vela XE-1/HD 77581	–	8.95	Rapid X-ray fluctuations	B0	–
SMC X-1/Sk 160	–	3.9	–	B super-giant	Extragalactic

4.5 The physics of the X-ray emission

The mass of a member of a binary system is found from measurements of the orbital parameters, which can be combined to give a ratio of masses known as the mass function. For Cyg X-1 the orbit is known from the period and from the Doppler shift of visible spectral lines; the mass function contains only one unknown, which is the inclination i of the orbit. The mass function is

$$M_X^3 \sin^3 i/(M_X+M_{HD})^2 = 0.23 M_\odot \qquad (4.2)$$

where M_X and M_{HD} are respectively the masses of the condensed star and the star HDE 226868, and $M_\odot$ is the mass of the Sun. For a given value of M_{HD}, the smallest value for M_X is obtained by setting $\sin i = 1$.

Estimates of the mass of HDE 226868 are based on its spectral type. Assuming that the mass loss has no effect on this estimation, the mass is believed to be around $25 M_\odot$, giving $M_X \approx 6 M_\odot$. If the condensed star is to be a neutron star, its mass cannot be greater than $1.7 M_\odot$, which would only be possible if M_{HD} were $3 M_\odot$ or less. The argument seems to be safely in favour of the 'black hole'. However, the conclusion still depends on the correctness of the identification of the star, and on the assumption that the relation between its luminosity and spectral type is unaffected by the mass transfer.

The other known binaries have mass functions which lead to lower masses for the condensed star. Cyg X-1 is the only one which seems to require a mass greater than the critical figure of $1.7 M_\odot$, and the only candidate for a 'black hole'. Even here the argument may have been oversimplified, since further dynamical components could be present, giving triple or more complex systems for which masses could not yet be determined.

4.5 The physics of the X-ray emission

The existence of regular pulsations in Her X-1 and Cen X-3 not only identifies the source star as a condensed rotating object, but it shows that there is an asymmetry in the object. As for the radio pulsars, the source is believed to be a strong permanent magnetic field, which channels and confines the flow to a zone at one or both poles. Unlike the radio pulsars, however, this is an inward flow under gravity, and not an outward flow due to electric fields.

The flow into a polar region follows a kind of funnel ending in an area only a few kilometres across on the surface. The enormous release of X-ray energy in this region, amounting to 10^{38} erg s^{-1}, creates a large outward radiation pressure, so that the inflow is halted above the surface. The X-rays are in fact emitted from this very hot patch of atmosphere.

The X-ray pulsars

Davidson (1973) has estimated that the hot region is no more than 200 m thick, with a mass per unit area about the same as in the Earth's atmosphere and a temperature of 6×10^7 K. It is astonishing to find that an easily detectable thermal source on a distant star can be as small as Hyde Park in London, and no taller than the Post Office Tower (a comparison due to F. D. Kahn). The 'pulsation' of the X-rays is supposed to be due to the periodic appearance of this bright patch as the star rotates.

The periodic variation of light output of the bright companion star cannot be due to an occultation, since the neutron star could only cover a negligibly small part of the stellar disc. It is in fact not a decrease, but an increase of brightness when the neutron star is between the observer and the stellar disc. This is easily explained as a local heating effect on the bright star due to the X-rays, and possibly also energetic particles, emitted by the neutron star.

A far more difficult problem is posed by the 35-day variation in the pulsed X-ray emission from Her X-1. Throughout the period the 1.7-day orbital variation of the light from HZ Her continues practically unchanged. If this 1.7-day variation is due to X-ray heating, how can the heating remain unchanged when the pulsed emission is varying by more than an order of magnitude through the 35-day cycle? Many solutions have been proposed; for example it has been suggested that the heating is primarily due to particles instead of X-rays, or possibly due to lower energy X-rays (about 0.5 keV) than those detected in the satellite telescopes (several keV).

Working from the only available explanation of the length of the period itself, i.e. from precession in the neutron star, Lamb, Pethick & Pines (1973) developed a geometrical model which would explain the modulation of the pulsed X-rays. In this model the accretion process takes place via an accumulation disc, whose axis is parallel to the axis of the binary star orbit. If the magnetic pole of the neutron star is not aligned along its rotation axis, then the infall to the pole will be modulated at the rotation rate as the relative orientation of the polar magnetic field lines and the orbits of the particles in the accumulation disc change. Then, if the rotation axis precesses, possibly even only by a few degrees, a major change in the rate of infall might occur over the precession cycle.

Our understanding of the X-ray sources has advanced by large strides since the discovery of the X-ray pulsars, but it is obvious that the details of the pulsing process, and of its modulation in a 35-day cycle, are as yet poorly understood. We shall see that a similar situation appertains to the radio pulsars, where only basic mechanisms and configurations are understood, leaving many subtleties and even some clear periodicities without any good explanation.

Finally, it is still a matter of speculation whether the binary model can explain all the galactic X-ray sources, apart from the supernova sources. The main evidence in favour is that all the sources which can be studied in sufficient detail show fluctuations on a time scale less than 1 s. It seems that all sources are physically small, possibly as small as neutron stars. If they do not pulsate periodically, it is reasonable to suppose that this is explicable on geometrical grounds; for example the magnetic pole might be aligned with the rotation axis, or the 'hot spot' might be obscured by scattering in the accumulation disc. For the purposes of argument about populations (as in Chapter 19) we may assume that the evidence allows us to place all these sources in a single class of objects.

References

Davidson, K. (1973). *Nature Phys. Sci.* **246**, 1.
Giacconi, R. (1974). *Proc. 16th Int. Solvay Conf. on Physics, Brussels*, p. 27.
Giacconi, R., Gursky, H., Kellogg, E., Schreier, E. & Tananbaum, H. (1971). *Astrophys. J.* **167**, L67.
Hayakawa, S. & Matsouka, M. (1964). *Prog. theor. Phys. Suppl.* **30**, 204.
van den Heuvel, E. P. J. & de Loore, C. (1973). *Nature, Lond.* **245**, 117.
Lamb, F. K., Pethick, C. J. & Pines, D. (1973). *Astrophys. J.* **184**, 271.
Zel'dovich, Ya. B. & Guseynov, O. K. (1965). *Astrophys. J.* **144**, 840.

5
The internal structure of neutron stars

The first suggestion that there might be a stable state of matter at extremely high densities, where the principal component would be a degenerate neutron fluid, was made by L. Landau in 1932. A possible application in astrophysics was suggested soon after by Baade & Zwicky, who in 1934 proposed that supernova explosions might leave a core which would condense into a neutron star. There was at this time no information on the equation of state of condensed neutrons which could lead to accurate estimates of the allowable range of masses, central densities, and radii of stable neutron stars.

A simple equation of state was applied by Oppenheimer & Volkoff in 1939, but detailed calculations awaited experimental measurements of neutron–neutron interactions at energies of 100 MeV and more. The stimulation provided by the discovery of the pulsars, and their identification as neutron stars, led to a spate of theoretical work which has provided a surprisingly detailed description of the structure of stars over the allowable range of masses.

5.1 Temperature

In contrast to more normal stars, in which the material is supported by a kinetic pressure which depends on temperature, almost the whole of a neutron star is effectively cold. The Fermi energy of degenerate heavy particles is over 50 MeV, so that a neutron gas is still 'cold' if its temperature is, say 10^{10} K, corresponding to a particle energy of 1 MeV. It is the zero-temperature energy of the particles, rather than thermal energy, which supports the star and prevents its further collapse. The equation of state which is required is solely a formulation of pressure as a function of density. At each density we must seek the state of matter which represents the lowest potential energy.

The outermost layers are not degenerate, and their temperature is of some interest in connection with the X-ray pulsars (Chapter 4). Cooling of the whole star is rapid after a possible formation at high temperature. Fig. 5.1 shows the results of computations by Tsuruta, Canuto, Lodenquai & Ruderman (1972). The cooling is accelerated by strong magnetic

5.1 Temperature

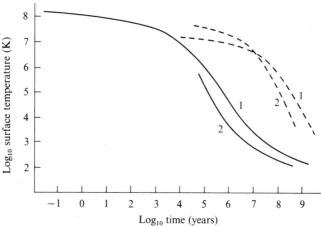

Fig. 5.1. Cooling curves for neutron stars. Curve 1 is for zero magnetic field, and 2 is for a field of 1.1×10^{12} gauss. The solid curves include the effect of superfluidity. (After Tsuruta et al., 1972.)

fields, which modify the opacity of the interior, and by superfluidity in the neutron fluid. The four curves show the cooling curves for zero field and a field of 1.1×10^{12} gauss (1 and 2 respectively), with and without the effects of superfluidity (solid and dashed lines respectively). The initial temperature is assumed to be greater than 10^7 K, when the cooling is rapid and mainly due to neutrino emission. As the temperature falls towards 10^6 K the dominant process becomes photon emission.

The conductivity is anisotropic in the presence of a strong magnetic field, since in practice the quantum energy corresponding to the cyclotron frequency is greater than the thermal energy, i.e.

$$\frac{h}{2\pi} \frac{eH}{mc} > kT$$

where h is Planck's constant. The magnetic poles may then have a larger conductivity than the equator by a ratio of between 70 and 90, the same ratio as for example between the conductivities of copper and asbestos. The poles will then be up to three times hotter than the equator (Smoluchowski, 1972). The actual temperatures are nevertheless very much smaller than the observed polar temperatures of X-ray pulsars. Except for the first few years of its existence, we can assume that the interior temperatures of a neutron star do not affect its structure. The surface temperature is expected to be of the order 10^3 to 10^4 K at a typical pulsar age of 10^6 years. This has no consequences in the structure of the surface, and no detectable X-ray emission can be expected. The very low

Internal structure of neutron stars

temperature may, however, be of importance for the theory of the pulsar magnetosphere, since it may not be correct to assume that thermal electron emission is sufficient to supply the continuous outward flow of charge required in the simple theory (Chapter 6).

5.2 The outer layers

Proceeding from the outer layer to the centre of a neutron star involves covering a range of about nine orders of magnitude in density, from 10^6 to 10^{15} g cm^{-3}. It is convenient to follow the structure in ranges of density starting at the surface.

At the surface the density of 10^6 g cm^{-3} is similar to the density of white dwarf stars. At this comparatively low density the total energy is a minimum when the material is condensed into heavy nuclei, primarily ^{56}Fe. Iron compressed to densities over about 10^4 g cm^{-3} will be fully ionised; the energy for this cold ionisation comes from the work done in the compression. The electron gas in the crust is degenerate.

The nuclei find a minimum energy when they are arranged in a crystalline lattice. The internuclear forces in this lattice are large, since the nuclei appear as bare charges. Melting would only take place if the thermal energy of the particles approached a small fraction, say 1%, of the energy of the Coulomb interaction, i.e. the melting temperature would be given by

$$100\, kT \approx \frac{e^2 z^2}{r}$$

where r is the average distance between the nuclei. This gives a melting temperature in excess of 10^8 K for the crust, well above the temperatures which are possible within a very short time after formation of the star.

In a typical neutron star the solid crust has a thickness of some hundreds of metres. We shall be concerned with the strength of the crystalline material in Chapter 7, where we shall consider the sudden changes in moment of inertia which are revealed by the timing observations.

As the density increases, nuclei with larger mass appear in the crust. This occurs because of the increasingly high Fermi energy of the electron field, which eventually becomes high enough to force electrons to combine with protons within the nuclei, forming nuclei with unusually high numbers of neutrons. Baym, Pethick & Sutherland (1971) have listed the most stable nuclei to be expected as a function of density up to 4×10^{11} g cm^{-3} (Table 5.1). The sequence soon includes species unstable in the laboratory, and ends with ^{118}Kr, a remarkable nucleus with eighty-

5.3 The neutron drip point

two neutrons and thirty-six protons. The ratio z/A of protons to total number of nucleons falls from 0.46 for ^{56}Fe to 0.305 for ^{118}Kr.

TABLE 5.1 *Stable nuclei in the neutron star crust*

Nucleus	z/A	ρ (g cm^{-3})	μ_e (MeV)
^{56}Fe	0.4643	8.1×10^6	0.95
^{62}Ni	0.4516	2.7×10^8	2.6
^{82}Ge	0.4048	8.2×10^9	7.7
^{80}Zn	0.3750	4.8×10^{10}	13.6
^{124}Mo	0.3387	1.9×10^{11}	20.2
^{118}Kr	0.3051	4.3×10^{11}	26.2

After Baym *et al.* (1971).
ρ is the maximum density at which each nuclear species is present.
μ_e is the electron chemical potential at that density.

This process of 'neutronisation' proceeds only as far as a density of 4.3×10^{11} g cm^{-3}. The precise sequence of these unfamiliar processes is unimportant, even though it must be more complex than the mere addition of electrons into the nuclei, since this would produce only species of constant mass. It is an assumption that the equilibrium states represented by the heavy nuclei will in fact be established in a reasonably short time: the detailed processes by which this would occur have not been worked out.

5.3 The neutron drip point

Up to a density of 4.3×10^{11} g cm^{-3} there are almost no neutrons outside the nuclei. Above this density the lowest energy state involves nuclei embedded in a fluid both of electrons and neutrons.

This transition occurs when the nuclei that should be in equilibrium are no longer stable, tending to release neutrons. A new regime is established, with nuclei of atomic number up to about 40 in a fluid with increasing neutron density. This regime persists up to a density of about 10^{14} g cm^{-3}, when the separate nuclei lose their identity, and the fluid becomes a continuous neutron fluid with electrons and protons as minor constituents. The details of the 'neutron drip' regime have been difficult to establish; nevertheless the equation of state is known fairly well, and it is certain that there is a degenerate neutron fluid present at densities above 4.3×10^{11} g cm^{-3}, which is known as the 'neutron drip point'. These results are sufficient for the description of neutron stars in any observational context.

5.4 The neutron fluid: superfluidity

The equation of state of the neutron fluid becomes progressively more uncertain at densities above 10^{14} g cm^{-3}; it is not known at all for densities above 10^{16} g cm^{-3}. At this density, which is the maximum density thought to be possible at the centre of the most massive neutron stars, the interparticle spacing is 0.7 fm, which is close to the neutron 'core radius' of 0.5 fm. Laboratory measurements of neutron interactions at energies approaching 1 GeV would be needed for measurement of nucleonic forces at such short distances.

Nevertheless, the uncertainties in the equation of state apply only to a small part of the total mass of a neutron star, so that the overall computations relating radius to mass are not much affected.

The degenerate neutron fluid is a superfluid, i.e. its viscosity is zero, or very nearly zero. This is not to be expected in a simple degenerate Fermi system: indeed the electron gas does not behave correspondingly as a superconductor. The difference is that neutrons with energy close to the Fermi energy attract each other, forming pairs. The energy spectrum then shows a gap of order 1 MeV wide, which prevents the re-distribution of energies which must occur for viscosity to be effective. The same effect leads to superconductivity in the proton fluid. This theory is the Bardeen, Cooper & Schriffer (BCS) theory of superconductivity and superfluidity.

The superfluidity of the neutrons is important in the explanation of irregularities of pulsar timing (Chapter 7).

5.5 Solidification of the core and formation of hyperons

There is good experimental evidence from timing observations (Chapter 7) that the Vela Pulsar contains a solid core in the centre of the neutron fluid. Theory gives little guidance to the circumstances in which the central part will solidify, since there is very little energy difference between the fluid and solid states. The difficulty is that the energy of the solid depends not only on the spherically symmetrical forces between two neutrons, but also on the higher-order terms involving neutron spin which depend on the relative position and orientation of adjacent members of a crystal lattice. Solidification is more likely at increased density, but estimates of the actual density at which it occurs vary over a wide range centred roughly on 10^{15} g cm^{-3}.

A solid neutron core will, like the neutron liquid, contain a proportion of electrons and protons as degenerate liquids. Its electrical conductivity is therefore very high.

At the highest densities, where the Fermi energy reaches 1 GeV, it is likely that fundamental particles with higher mass will be formed. The

5.6 Neutron star models

uncertainties in our knowledge of forces between the hyperons do not allow firm conclusions as to the density where they will first be formed, but it is probably close to 10^{15} g cm^{-3}. The first hyperon to appear is the Σ^-, followed by Λ and Σ^0.

The presence of hyperons seems to have little effect on the equation of state of the core material, and there may be no way of deducing their presence from any observable property of pulsars.

5.6 Neutron star models

The equations of state for the series of density regions described above can now be put together to make models of neutron stars with various masses. The results have been presented by Cameron & Canuto (1973).

Figure 5.2 shows the central density as a function of total mass for a wide range of highly condensed stars. The two curves, for a solid and for a

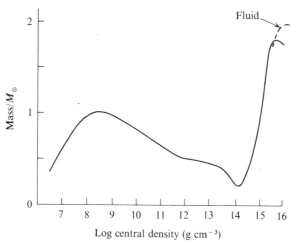

Fig. 5.2. Central density as a function of total mass for a wide range of condensed stars. White dwarf stars are at the left of the diagram, and neutron stars at the right. (After Cameron & Canuto, 1974.)

liquid model, differ only slightly, as is to be expected from the small differences in the equations of state of solid and liquid. Only points on positively sloping portions of the curve represent stable states. At the left of the curve the stable states represent white dwarf stars, while neutron stars lie on the steeply rising curve at the right. The main conclusions are:

(i) the central density of neutron stars is between 2×10^{14} and 4×10^{15} g cm^{-3};
(ii) their mass lies between 0.15 and 1.7 solar masses.

Internal structure of neutron stars

The distribution of density with radius in neutron stars with masses from 0.5 to 1.5 solar masses is remarkably simple. Two examples are given in Fig. 5.3. Effectively all the mass is contained within a radius of 10 km, the heaviest stars having the smallest radii. The density does not vary rapidly, except in the outer crust. At the lowest mass the crust extends further out, to a radius of 50 km, and the boundary is more diffuse.

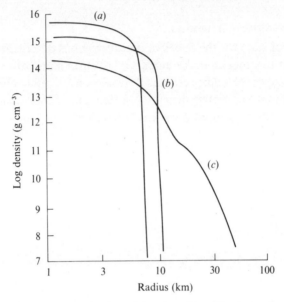

Fig. 5.3. Distribution of density with radius in neutron stars with three different masses (in units of the solar mass). (a) $1.4M_\odot$; (b) $0.5M_\odot$; (c) $0.1M_\odot$. (After Cameron & Canuto, 1974.)

It may be convenient to remember the simple picture of Fig. 5.4, which shows the various regions of a typical neutron star. The actual positions of the boundaries depend on the mass of the star.

5.7 The stability and the formation of neutron stars

The analysis of neutron star structure presented in this chapter has been concerned solely with the static situation, saying nothing about the possible ways in which the star may have been formed. In fact it is unlikely that neutron stars with the whole range of mass from 0.15 to 1.7 solar masses ($M_\odot$) will be formed. At the lower end the binding energy is less than the potential energy of the whole mass dispersed in the form of helium atoms, so that it is impossible for an isolated dispersed cloud of $0.15M_\odot$ to become a neutron star.

5.7 Stability and formation of neutron stars

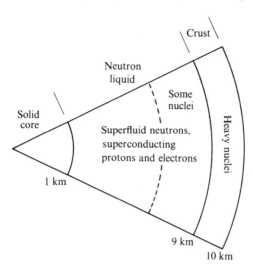

Fig. 5.4. Typical cross-section of a neutron star. Neutron stars with low mass may not have a central solid core, and their outer parts will be more extended.

Up to $1M_\odot$ there is a stable configuration as a white dwarf (Fig. 5.2), which acts as a barrier for a condensing object. Beyond $1.7M_\odot$ there is no stable configuration which will prevent a complete collapse under gravitation. Heavier masses are therefore expected to collapse into 'black holes', from which no radiation can emerge and which are only detectable through their gravitational field. A collapse to a neutron star must therefore involve less than $1.7M_\odot$, while if the mass is less than $1M_\odot$ the 'trapped' configuration of the white dwarf must be avoided. We shall see later that even the formation of a 'canonical' neutron star, with mass $1M_\odot$ and radius 10 km, involves the catastrophic event of a supernova explosion, in which the precise sequence of events is hard to follow. There is no guidance here as to whether masses greater or less than $1M_\odot$ are to be expected.

With all the uncertainties of formation and structure outlined in this chapter it is remarkable that the neutron star emerges from analysis as a closely defined object, with mass lying roughly in the range 0.5 to $1.5M_\odot$ and radius close to 10 km. The structure of the magnetosphere surrounding the condensed star also has some clearly determined characteristics, but it is at the same time a less predictable and more complex physical region, both theoretically and from observations.

43

References

Baade, W. & Zwicky, F. (1934). *Phys. Rev.* **45**, 138.
Baym, G., Pethick, C. & Sutherland, P. (1971). *Astrophys. J.* **170**, 299.
Cameron, A. G. W. & Canuto, V. (1974). *Proc. 16th Solvay Conf. on Physics, Brussels*, p. 221.
Landau, L. (1932). *Physik Zeits. Soviet union* **1**, 285.
Oppenheimer, J. R. & Volkoff, G. M. (1939). *Phys. Rev.* **55**, 374.
Smoluchowski, R. (1972). *Nature Phys. Sci.* **240**, 54.
Tsuruta, S., Canuto, V., Lodenquai, J. & Ruderman, M. (1972). *Astrophys. J.* **176**, 739.

6

The magnetosphere of neutron stars

When neutron stars existed only in theory, and there was little expectation that they could ever be observed, it was natural for theoretical work to be concentrated on the very remarkable solid-state physics of the interior. There was no expectation that an appreciable atmosphere could exist, since the star was expected to be cold and gravitational forces were expected to be so high that any atmosphere would be practically condensed on to the surface. The situation was completely transformed by Pacini's suggestion in 1967 that there might be a very strong magnetic field in neutron stars, and that they might be rotating so rapidly that the Lorentz force of this field would be of great importance. In fact, as we will see, this force completely dominates the magnetosphere, overwhelming gravity by many orders of magnitude.

When the Crab Pulsar was first observed to be slowing down, the same model immediately occurred to Gold (1968). He very easily fitted together an elementary model of a neutron star, and the observed rotation period, with the observed rate of slowing, and deduced that the Crab Pulsar contained 10^{49} ergs, and that it was losing 10^{38} erg s^{-1}. These figures fitted with Pacini's predictions in the following way.

According to classical electrodynamics, a magnetic dipole with moment $m_\perp$, rotating at angular velocity Ω about an axis perpendicular to the dipole, radiates a wave at angular frequency Ω with a total power $\frac{2}{3}m_\perp^2\Omega^4c^{-3}$. The energy supply must be the angular kinetic energy of the rotating body. Let the moment of inertia be I. Then

$$\frac{d}{dt}\tfrac{1}{2}I\Omega^2 = I\Omega\dot\Omega = -\tfrac{2}{3}m_\perp^2\Omega^4c^{-3}. \tag{6.1}$$

As we have seen in the previous chapter, the dimensions and hence the moment of inertia of a neutron star are known within reasonable limits, so that measurements of Ω and $\dot\Omega$ are all that is needed to determine the magnetic moment $m_\perp$. The result is that the surface magnetic field strength of the Crab Pulsar must be about 10^{12} gauss.

At first sight this field seems impossibly high. It could, however, easily be derived from the original magnetic field of a normal star before

Magnetosphere of neutron stars

collapse. If, for example, the Sun were to collapse to the size of a neutron star, conserving magnetic flux as the radius r shrinks, so that Br^2 remains constant, then the field strength B would increase by a factor 10^{10}. Polar fields of the order of 100 gauss are believed to be common in stars, and fields of several thousand gauss are known in some 'magnetic' stars. Although there seemed to be a ready explanation for the origin of the field, there was at first some doubt about the lifetime of such a strong field, since it might decay by ohmic dissipation in a lifetime short compared with that of pulsars. This proved not to be so, since the interior of neutron stars was seen to be superconducting.

Pacini's analysis of radiated energy at the rotation frequency was related to the component of the field perpendicular to the rotation axis. A dipole field aligned with the rotation axis would not radiate at all. However, as we shall see in the next section, there would be a magnetosphere of charged particles surrounding a rotating neutron star with a strong magnetic field. Part of this magnetosphere would be flowing continuously out in the form of a stellar wind, carrying angular momentum and rotational kinetic energy with it. A remarkable analysis by Goldreich & Julian, referred to later in this chapter, showed that the rate of loss of rotational kinetic energy by the wind expected in an aligned rotator with magnetic moment $m_\|$ was given by $\frac{2}{3}m_\|^2\Omega^4 c^{-3}$, exactly as for the dipole radiation except for the substitution of $m_\|$ for $m_\perp$. This apparent coincidence is discussed further at the end of this chapter.

6.1 Electrodynamics of the magnetosphere

The main theoretical problem of the pulsar magnetosphere can be stated very simply, although the solution proves to be very difficult. If a strongly magnetised sphere is rotated rapidly in a vacuum, the surrounding electromagnetic field is easily obtainable from classical theory. If individual test charges are introduced into this vacuum field, then their behaviour can be described in terms of the field. But if a considerable charge density can exist in the space around the sphere, then the field may be modified by the charges and a self-consistent solution is required. This self-consistent solution has not yet been obtained for the conditions known to obtain in a neutron star; it is even possible that no static solution exists, and it seems likely that the solution may depend critically on conditions at the surface of the star.

Before intense radio emission was observed from the pulsars, it was thought that neutron stars would have practically no atmospheres. A neutral atmosphere under gravitational equilibrium would have a scale

6.1 Electrodynamics of the magnetosphere

height h for neutral hydrogen, given by

$$h = \frac{kTr_0^2}{GMm} \tag{6.2}$$

where the neutron star has mass M and radius r_0, and m is the mass of a hydrogen atom. For the typical neutron star, and with a surface temperature $T = 10^6$ K, this density scale height is approximately 1 cm only, so that the atmosphere is practically non-existent. The necessary existence of an atmosphere of charged particles, which may best be thought of as an electrically generated magnetosphere, was first proved in a classic paper by Goldreich & Julian (1969).

The existence of the magnetosphere depends on the very large electric fields which would be generated by the magnetic field of the rotating star, if it were surrounded by a vacuum. Whatever the configuration of the magnetic field, dipolar or multipolar, aligned or skew with respect to the rotation axis, there will be an electric field immediately outside the surface of order $\Omega r_0 B/c$, where Ω is the angular velocity and B the magnetic field. For the Crab Pulsar the observed periodicity gives $\Omega \approx 200$, and B is known to be about 10^{12} gauss from the observed rate of energy release. The field is then of order 10^{12} V cm^{-1}. The total potential differences existing close to the star, for example between pole and equator for an axisymmetric dipole, would then be of order 10^{18} V. The field at the surface of the Crab Pulsar is strong enough to overcome any electrostatic binding forces, and electrons or protons will be emitted. They will attain a total energy of about 10^{18} V unless the field is modified by a build-up of charge.

The electric potentials that can be generated in this way are almost sufficient to accelerate charged particles to the highest known cosmic ray energies. Even when the potential is reduced by the establishment of charges in the magnetosphere, there may be sufficient potential for the pulsars to be a significant source of lower energy cosmic rays. By way of contrast, it is interesting to calculate the potential that can be generated by a rotating permanent magnet on a laboratory scale: for example, if a magnet with dimensions of order 10 cm and field 10^4 gauss is spun at 10^4 rad s^{-1}, only a few volts can be generated.

A comparison of gravitational and electrostatic forces on charges near the surface shows the overwhelming influence of the induced electric field. For the Crab Pulsar the ratio

$$\frac{GMm}{r^2} \bigg/ \frac{e\Omega B}{c} \tag{6.3}$$

Magnetosphere of neutron stars

for a proton is about 2×10^{-9} and for an electron 10^{-12}. Alternatively one may express the gravitational binding energy in terms of electron volts, to be compared with the potential differences generated across the pulsar surface. These binding energies for a typical pulsar are about 10^8 eV for a proton and 10^5 eV for an electron, many orders of magnitude less than the potential differences in the vacuum case. Gravitational forces are therefore completely negligible.

6.2 Axisymmetric dipole

Goldreich & Julian analysed the fields and charge densities expected to be built up in the magnetosphere, for the axisymmetric dipole. The magnetosphere is highly conducting along, but not perpendicular to, the magnetic field lines. This condition in the magnetosphere is very similar to the full conductivity of the stellar interior, where there can be no net electric field. The magnetosphere then seems to be an extension of the solid interior; in both regions the induced electric field is cancelled by a static field, so that

$$\mathbf{E} - \frac{1}{c}(\mathbf{\Omega} \times \mathbf{r}) \times \mathbf{B} = 0. \tag{6.4}$$

In this fully conducting situation there must be a charge density equal to $(1/4\pi)$ div $\mathbf{E}$; a simple analysis then shows that the difference in numbers of positive and negative charges is given by

$$n_- - n_+ = \frac{\mathbf{\Omega} \cdot \mathbf{B}}{2\pi e c}. \tag{6.5}$$

As a useful guide, the particle density is given approximately by

$$n_- - n_+ = 7 \times 10^{-2} \frac{B_z}{P} \text{ cm}^{-3} \tag{6.6}$$

where B_z is the axial component of the field in gauss, and P is the period in seconds.

If the conducting sphere were surrounded by a vacuum, there would be an electric field of order $\Omega r_0 B/c$ immediately above the fully conducting surface, with a component perpendicular to the surface. This means that there must be a corresponding surface charge. When there is a conducting magnetosphere a charge density is built up outside the star in the same way as it is in the interior. In the perfectly conducting sphere there is no component of electric field along the magnetic field, i.e. $\mathbf{E} \cdot \mathbf{B} = 0$, which is the same condition for the field in the magnetosphere. The lack of

6.2 Axisymmetric dipole

conductivity perpendicular to the magnetic field within the magnetosphere therefore has no effect, and the field pattern is continuous from the interior to the magnetosphere. The net charge density is again given by (6.5), the main difference being that there may be complete charge separation in the magnetosphere instead of a small excess of one sign of charge.

This magnetosphere of fields and charges is co-rotating with the neutron star, locked firmly to it by the magnetic field. The electrodynamic condition for this locking is that the Alfven velocity (for magnetodynamic waves) is less than the co-rotation velocity $\boldsymbol{\Omega} \times \mathbf{r}$; if this were not so then any instability in the field pattern would fall behind the rotation and the pattern would be destroyed. This appears not to be an important restriction, and co-rotation can continue until the velocity $\boldsymbol{\Omega} \times \mathbf{r}$ approaches the velocity of light. The surface where $\boldsymbol{\Omega} \times \mathbf{r}$ reaches c is a cylinder, known as the velocity-of-light cylinder. At and beyond this cylinder the co-rotation must break down. In fact (6.4) and (6.5) must be modified even inside the velocity-of-light cylinder, as discussed later.

The pattern derived by Goldreich & Julian is shown in Fig. 6.1. The closed field lines are those which originate on the surface beyond an angular distance θ_0 of the poles, given by

$$\sin \theta_0 = \left(\frac{\Omega r_0}{c}\right)^{1/2}. \tag{6.7}$$

The extreme closed field lines meet the velocity-of-light cylinder tangentially, enveloping a completely closed region of co-rotating magnetosphere. Lines originating close to the poles cross the velocity-of-light cylinder, entering a region which cannot co-rotate. The zone from which particles may leave the magnetosphere is known as the 'wind zone', although it should be noted that the trajectory of charged particles can no longer be restricted to following the open magnetic field lines as they approach the velocity-of-light cylinder.

The induced electric field has the same sign at each pole, so that the open field lines have the same potential difference between the surface and outside the velocity-of-light cylinder. Particles of only one sign cannot flow continuously, so the star builds up such a potential that the flow is divided as shown into proton and electron streams. The polar region will therefore have two zones, named by Sturrock (1971) the EPZ and the PPZ, where only electrons or only protons stream out from the surface.

Although it is clear that the magnetospheric electrons and protons are constrained to follow the magnetic field lines closely, at least until they

Magnetosphere of neutron stars

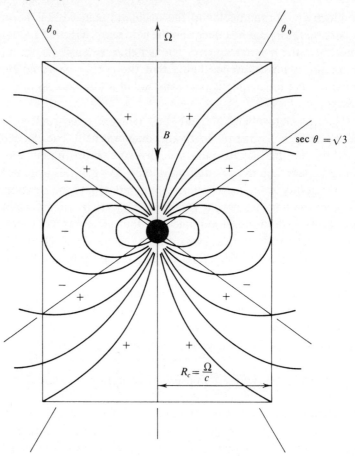

Fig. 6.1. The Goldreich & Julian magnetosphere. The polar angle θ_0 divides the open and closed field lines. The positive and negative regions divide at $\sec\theta = \sqrt{3}$. Co-rotation occurs out to the velocity-of-light cylinder at $R_c = \Omega/c$.

are near the velocity-of-light cylinder, there is no agreement about the distribution of the accelerating component of the electric field along the magnetic field lines. There is a possibility, discussed in Section 6.5, that a vacuum gap may develop near the surface, inside which the electric field may be very large, so that particles obtain most of their energy as they cross the gap in a discharge process. Immediately outside the gap, however, the field would be small, as it would be close to the surface for a continuous magnetosphere without a gap. In either case it seems that a large part of the acceleration in the polar field lines must take place within a distance comparable to the radius of the polar cap.

6.4 Complications and instabilities

The particles accelerated in the polar cap region will probably reach energies sufficient for a cascade process of pair creation to take place (Sturrock, 1971). The charged particles are constrained to follow a curved path, and the consequent curvature radiation (Chapter 15) can reach gamma-ray energies. The gamma-rays in turn encounter the strong magnetic field, and create pairs of electrons and positrons. These particles may again be accelerated, radiating gamma-rays, which create more pairs. This cascade process may be important as a source of charged particles.

6.3 The oblique rotator

We have already noted that the electrodynamic braking force is similar for the aligned and transverse rotating dipoles, even though the former requires the existence of a magnetosphere for braking to occur. There are in fact many similarities between the magnetospheres generated in the two cases.

Mestel (1971) showed that many of the Goldreich–Julian arguments developed for the aligned rotator applied equally to the oblique rotator. As before, charges will be pulled out of the star until the electric field component along **B** is near zero. As before, the charge density is of order ΩB, but it varies with longitude instead of having the circular symmetry of the aligned case. There is again a region of closed field lines, and an open polar region. The pattern of field lines at the velocity-of-light cylinder is, however, fluctuating as the star rotates: this represents the oscillating field of the rotating dipole, modified by the charge density and currents in the outer part of the magnetosphere. *In vacuo*, the magnetic field changes from radial at the surface to toroidal beyond the velocity-of-light cylinder, and the radial dependence changes from r^{-3} to r^{-1}.

The magnetic field still plays the dominant role, being substantially unmodified by the particles except close to the velocity-of-light cylinder. The flow of energy across the velocity-of-light cylinder is reasonably well represented by the flow of magnetic energy $(B^2/8\pi)4\pi R_c^2 c$, although there must also be a component of particle energy.

6.4 Complications and instabilities

Beyond the general description given so far there are many complications. An outstanding question is raised by the assumption that $\mathbf{E} \cdot \mathbf{B} = 0$, i.e. that there is no electric field along magnetic field lines. There must in fact be some such field to accelerate the particles. Provided that the inertial mass of the particles is small, only a very small field component need exist. But when the particles acquire a very high energy, and even

Magnetosphere of neutron stars

more when they are forced into co-rotation near the velocity-of-light cylinder, the field must change. There is already an over-simplification of this sort in (6.5), since the charge density should properly be written

$$\rho = \frac{\nabla \cdot \mathbf{E}}{4\pi} = -\frac{\Omega}{2\pi c}[\mathbf{B} - \tfrac{1}{2}\mathbf{r} \times (\nabla \times \mathbf{B})] \tag{6.8}$$

where the second term becomes important when the velocities approach c.

A further indication of trouble in the simple solution is given by Holloway (1973), who pointed out a remarkable form of instability in the fully charge-separated plasma. If for some reason a region in the magnetosphere is emptied of charges, a potential difference is set up which prevents charges re-entering it. In fact it will probably re-fill by processes of diffusion or turbulence, but only slowly. The most remarkable aspect, however, is that the charged magnetosphere on either side of a hollow slab will move at different speeds, so that an extended toroidal slab could separate regions which were not co-rotating properly with the neutron star, but gaining or losing in rotation rate.

6.5 Ruderman's whiskers and the work function

The Goldreich–Julian theory has been developed on the assumption that charges of either sign can be freely emitted at the surface, so that there is no discontinuity in electric potential along the magnetic field lines. This would be a safe assumption if the work function for charge emission at the surface were, as in normal material, only a few volts. The crystalline state of the surface material must, however, be profoundly altered by the intense magnetic field. Ruderman (1972) showed that the crystal lattice becomes very anisotropic, with a fibrous structure often referred to as 'Ruderman's whiskers'. The binding energy of electrons in this structure may easily reach 10 kV, which is too large for field emission to provide the necessary flow out into the magnetosphere.

If only protons are emitted, and the electron zones are cut off, the neutron star would soon charge up negatively and all particle emission would stop. There is, however, a way of providing a copious flow of charged particles without much surface emission, through the mechanism of pair production. The steady-state condition as envisaged by Ruderman & Sutherland (1975) is shown in Fig. 6.2. A large part of the induced electric potential in the magnetosphere is developed across a vacuum gap, possibly about 100 m thick, immediately above the star surface. If a single electron is released from the surface by field emission, it is immediately accelerated to very high energy, and in the presence of the intense

6.6 Energy flow from the pulsar

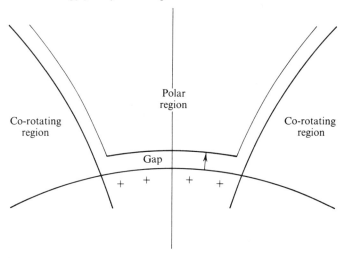

Fig. 6.2. Polar gap. The neutron star surface is positively charged below the vacuum gap, and a spark discharge can cross the gap as shown by the arrow. (After Ruderman & Sutherland, 1975.)

magnetic field it emits high-energy gamma-rays. The gamma-ray photons have sufficient energy to produce electron–positron pairs, which are themselves accelerated, forming a cascade of particles and gamma-rays. The upper surface of the vacuum slab then becomes the effective emitting surface. The emission may be concentrated at one or more points in the surface, like a spark discharge, and these points may move about.

This theory not only saves the simple magnetospheric model by providing sufficient charged particles: it also provides new parameters to fit to the more complex phenomena such as pulse drifting and the irregular moding and nulling described in Chapter 9.

6.6 Energy flow from the pulsar

It has already been remarked that the energy flow from a rotating neutron star is made up of two parts, the electromagnetic radiation and the outflow of particles. The remarkably close relation between these two can now be seen as a consequence of the physical conditions at the velocity-of-light cylinder.

The electromagnetic radiation of the perpendicular dipole can be evaluated roughly as follows. The field pattern at the cylinder can be regarded as flowing out with velocity c over an area of order $4\pi R_c^2$. The energy density is $B^2/8\pi$, so that the radiated power is $\frac{1}{2}B^2 R_c^2 c$. Since the surface field B_0 at radius r_0 is related to B by

$$B = B_0 \frac{r_0^3}{R_c^3}$$

we find that the radiated power is approximately $(B_0 r^3)^2 \Omega^4 c^{-3}$, which is comparable with the more precise expression $\frac{2}{3} m_\parallel^2 \Omega^4 c^{-3}$ obtained from classical theory. Similarly the particle energy flow crosses the velocity-of-light cylinder with velocity close to c, so that the total energy flow is

$$\frac{d\omega}{dt} = \left(\frac{B^2}{8\pi} + \rho\right) 4\pi R_c^2 c$$

where ρ is the particle energy density.

The reason for the apparent coincidence is now seen to be the interchange between $B^2/8\pi$ and ρ for the two cases. Evidently the field in the aligned case induces a stellar wind with energy density equal to the magnetic field density. The nature of the approximations in the expression for energy loss in both the aligned and perpendicular cases is now obvious: both depend on the almost unknown conditions at the velocity-of-light cylinder. Measurements of the rotational slowdown show that the rate of energy loss in the Crab Pulsar is not exactly proportional to Ω^4; the reason must evidently be related to the breakdown of the simple field configurations at the velocity-of-light cylinder.

References

Gold, T. (1968). *Nature, Lond.* **218**, 731.
Goldreich, P. & Julian, W. H. (1969). *Astrophys. J.* **157**, 869.
Holloway, N. J. (1973). *Nature Phys. Sci.* **246**, 6.
Mestel, L. (1971). *Nature Phys. Sci.* **233**, 149.
Pacini, F. (1967). *Nature, Lond.* **216**, 567.
Ruderman, M. (1972). *IAU Symposium No. 53.* (Dordrecht: D. Reidel.)
Ruderman, M. A. & Sutherland, P. G. (1975). *Astrophys. J.* **196**, 51.
Sturrock, P. A. (1971). *Astrophys. J.* **164**, 529.

7
Pulse timing

Astrophysics provides many examples of rotating and orbiting bodies whose periods of rotation and revolution can be determined with great accuracy. Within the Solar System the orbital motion of the planets can be timed to a small fraction of a second, while the rotation of the Earth is used as a clock which is reliable to about 1 part in 10^8 per day. Outside the Earth there is, however, no other clock with a precision approaching that of pulsar rotation.

The arrival times of the radio pulses from the pulsars are easy to study, and a surprising amount can be learned from them. Not only do they provide information on the nature of the pulsed radio source, they also can give an accurate position for the source, and they can give information on the propagation of the pulses through the interstellar medium. All three kinds of information were noted by Hewish and his collaborators in the discovery paper of 1968. They showed that the shortness of the pulses, and their short and precise periodicity, implied that the source was small, and that it might be a rotating neutron star. They showed also that the arrival time was varying due to the Earth's motion round the Sun; this annual variation implied that the source lay outside the Solar System, and they were able to obtain the celestial co-ordinates of the source from the amplitude and phase of the annual variation. Finally they showed that the arrival time of a single pulse depended on radio frequency; this dispersion effect was found to be in accord with the effect of a long journey through the ionised gas of interstellar space.

7.1 Pulsar positions and the barycentric correction

Since the time of Roemer, who made observations of the motion of Jupiter's satellites when the Earth was at different positions in its orbit, it has been known that light takes about $8\frac{1}{2}$ minutes to travel from the Sun to the Earth. Pulses from a pulsar lying in the plane of the ecliptic will therefore arrive earlier at the Earth than at the Sun when the Earth is closest to the pulsar, i.e. when it is at the same heliocentric longitude. Six months later the pulses will arrive late by the same amount. Assuming for simplicity that the Earth's orbit is circular and centred on the Sun, the

Pulse timing

delay t_c is given by

$$t_c = A \cos(\omega t - \beta) \cos \lambda \tag{7.1}$$

where A is the light travel time from Sun to Earth, ω is the angular velocity of the Earth in its orbit, and β, λ are the ecliptic longitude and latitude of the pulsar (Fig. 7.1a).

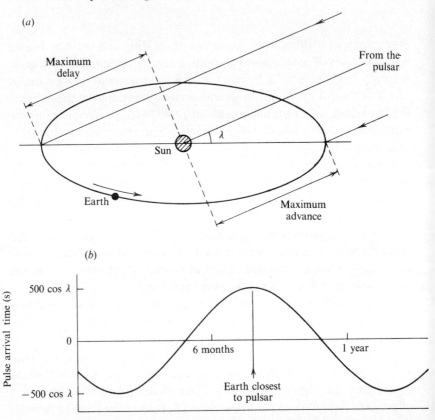

Fig. 7.1. (a) The annual variation in pulse arrival time due to the Earth's orbital motion round the Sun. (b) The amplitude of the variation is 500 cos λ s, where λ is the ecliptic latitude of the pulsar. The phase of the sinusoid is used to determine ecliptic longitude.

The observed arrival times of pulses emitted at equal time intervals throughout the year will therefore show a sinusoidal variation as in Fig. 7.1(b), where the phase and amplitude respectively give the heliocentric longitude and latitude of the pulsar. The accuracy of the positions is greatest near the pole of the ecliptic, since the ecliptic latitude λ is poorly determined near $\lambda = 0$. An error in co-ordinates $\delta\lambda$, $\delta\beta$ gives rise to

7.1 Pulsar positions and the barycentric correction

periodic timing errors

$$\delta t_c = -A\,\delta\lambda\,\cos(\omega t - \beta)\sin\lambda + A\,\delta\beta\,\sin(\omega t - \beta)\cos\lambda \qquad (7.2)$$

so that a latitude error $\delta\lambda$ is not detectable at $\lambda = 0$, whereas the angular error in longitude ($\delta\beta\cos\lambda$) is not a function of latitude.

Positions obtained from the variation of pulse arrival times through a year are remarkably accurate. Typically a pulsar with a period of about 1 s can give a point on the timing curve with an accuracy of about 2 ms. At least four such observations through 1 year are needed to find the source position, but the result is a position accurate to the order of 1 arc second. This is sufficiently accurate for any attempts at identification with unusual visible or X-ray objects, and positions have therefore been obtained from timing observations for some tens of pulsars. The application of the method requires consideration of the following details (Hunt, 1971);

(i) The pulsar period lengthens during the observation, adding a uniform slope to the curve of Fig. 7.1(b). (Determination of this slope may, of course, be a prime objective of the observations, since it gives a measure of pulsar age.)

(ii) Rotation of the Earth introduces a variable time delay up to the transit time over one Earth radius.

(iii) The Earth's orbit is elliptical, not circular.

(iv) The Sun moves in relation to the centre of inertia of the Solar System, known as the barycentre. This motion depends on the orbital motion of the planets, mainly the massive planet Jupiter; it is sufficiently large to place the barycentre just outside the surface of the Sun.

(v) The gravitational potential of the Earth differs from the potential at a large distance from the Sun: furthermore it varies annually through the ellipticity of the Earth's orbit. General Relativity then requires a careful definition of the mean reference clock, and predicts a small annual variation of clock rate.

(vi) The effective receiver frequency varies through the year due to the Doppler effect of the Earth's motion. Since the arrival time depends on frequency, due to dispersion in the interstellar medium, a correction may be needed for timing observations of pulsars with high dispersion measure.

In practice the expected arrival times are computed using an assumed position β, λ and an ephemeris giving the vector distance $\mathbf{r}_{ob}$ from the observer to the barycentre. This vector distance is the sum of the three vectors $\mathbf{r}_{oc}$ from the observer to the centre of the Earth, $\mathbf{r}_{cs}$ from the centre of the Earth to the centre of the Sun, and $\mathbf{r}_{sb}$ from the centre of the Sun to

Pulse timing

the barycentre:

$$\mathbf{r}_{ob} = \mathbf{r}_{oc} + \mathbf{r}_{cs} + \mathbf{r}_{sb}. \tag{7.3}$$

Then the time t_c to be added to the barycentric observation time is

$$t_c = -\frac{1}{c}\mathbf{r}_{ob} \cdot \mathbf{s} \tag{7.4}$$

where $\mathbf{s}$ is the position vector of the source at β, λ.

The three components of $\mathbf{r}_{ob}$ are obtained separately. The vector $\mathbf{r}_{cs}$ is available in terms of the astronomical unit; in practice it may be necessary to combine two ephemerides which give respectively the motion of the barycentre of the Earth–Moon system and the motion of the Earth within that system. The astronomical unit itself has been well determined as a light travel time from planetary radar. The barycentric correction $\mathbf{r}_{sb}$ is obtained from the vector positions $\mathbf{r}_i$ and masses m_i of the planets:

$$\mathbf{r}_{sb} = \frac{1}{1+\Sigma\dfrac{1}{m_i}} \Sigma \frac{1}{m_i}\mathbf{r}_i. \tag{7.5}$$

The Earth radius correction $\mathbf{r}_{oc}$ may be computed directly as a light travel time, since it depends only on the radius of the Earth at the observatory and the source elevation E. At the mean radius of the Earth the time correction is $21.2 \sin E$ ms.

The computations of all three components of the correction to the barycentre can be made to an accuracy of a few microseconds.

7.2 The relativistic correction

The relation between time intervals Δt_E measured at the Earth and the corresponding intervals Δt_∞ measured with an identical clock at an infinite distance from the Sun is, according to the Schwarzchild metric:

$$\Delta t_E = \Delta t_\infty \left(1 - \frac{GM}{c^2 r} - \frac{v^2}{c^2}\right)^{1/2} \tag{7.6}$$

where the observer is moving with velocity v at distance r from the Sun. The average of this difference is removed from consideration by defining the standard clock as a clock situated at the barycentre but running at the rate appropriate to the mean distance of the Earth from the Sun, i.e. the semi-major axis of the elliptical orbit. The variable terms are due to variations of r and v round the orbit.

The annual variation of the relativistic term consists of a sinusoid together with harmonic terms. The first term has a peak-to-peak amplitude of 3.4 ms, so that it is easily detectable in timing observations of the

7.3 Periods, period changes and pulsar positions

Crab Pulsar. It is not measurable in other pulsars whose positions are not known independently of the timing observations, since it has the same form as a small error in the position of the pulsar (about 1 arc second, depending on the ecliptic latitude). The second harmonic, with a period of 6 months, has a peak-to-peak amplitude of 28 μs; this is hard to detect even in the Crab Pulsar, since it is masked by other irregularities which apparently originate within the pulsar.

7.3 Periods, period changes and pulsar positions

All known pulsars, except the X-ray pulsars, have the basic characteristic of an intrinsically precise period, modulated only by a slow monotonic increase in period. The change of period of the Crab Pulsar can be detected within a few hours, and the change for the Vela Pulsar within days, but generally the rate of change is so small that it can only be determined from observations over a period of 1 year. Furthermore, accurate positions of pulsars are usually only available from the timing observations themselves, so that an apparent change of period over a short observing time may be due only to an error in the assumed pulsar position.

After many unsuccessful attempts at identifications with visible objects or X-ray sources, it now seems unprofitable to expend much observing time on obtaining accurate positions for pulsars except for the measurement of proper motion (Manchester, Taylor & Van, 1974, see Section 7.9). Timing observations are usually directed instead to obtaining the rate of change of period $\dot{P}$, which is used in finding the 'age' of the pulsars from the ratio $P/\dot{P}$. This can conveniently be done by measuring the period P at two epochs close to one year apart. If the interval is not exactly a whole year, it is possible to use an approximate position for correcting to a precise interval of one year; preferably, a series of observations over several days at each epoch can be used to extrapolate to the correct interval. Most of the ages discussed in Chapter 20 have been obtained in this way (Lyne, Ritchings & Smith, 1975).

The outstanding result from extended timing measurements is that the arrival times of the pulses are astonishingly regular. On a short time scale, i.e. within a few days, the arrival time shows only a small jitter from pulse to pulse, which averages out over a few hundred pulses. Observations of the arrival time smoothed in this way and continued over months or years usually show only the geometric effect of the barycentric correction, and the regular 'spin-down' represented by the first differential ($\dot{P}$) of the period. Discontinuities in arrival time, period (P), or its differential ($\dot{P}$) have been observed only in a few pulsars, notably the Crab Pulsar and the

Pulse timing

Vela Pulsar. The second differential $\ddot{P}$ has only been observed in the Crab Pulsar. Only one pulsar has shown the sinusoidal modulation of pulse arrival time expected from orbital motion in a binary system (although this was at one time suspected for the Crab Pulsar). The main conclusion is simple: the pulse timing is determined solely by the smooth rotation of the neutron star, with the source firmly locked to the inertial system of the star. Further, it is clear that the radio pulsars are not generally members of close binary systems, as are the X-ray pulsars Her X-1 and Cen X-3. This is of considerable importance in discussions of the life history of pulsars (Chapter 20).

We turn now to the exceptions to the general rule of such smooth and featureless behaviour.

7.4 The Crab Pulsar

This pulsar, PSR 0532+21, is the youngest known pulsar, with the shortest known period. Accordingly its rotational period can be timed with the greatest precision, and its changes in period can be most easily measured.

On a time scale of some days, the pulsar shows a remarkable uniformity of rotation rate. There is little difference between the results obtained by radio and optical methods, except that individual optical results tend to be more precise, while radio results are more easily obtained consistently over a long period. As an example of short-term precision, we note an observation by Pfleiderer (1971), timing radio pulses on 405 MHz over a period of 33.5 hours. A single timing observation represented an averaging over 5 minutes. The timing was continued for two whole days, from rising to setting of the Crab Nebula. Allowing for a linear change of period, the pulse arrival times were found to be consistent within 5 μs during one day, and within 15 μs between the two days. For comparison, 1° of rotation of this pulsar corresponds to 90 μs; the location of the source centre referred to the surface of the pulsar was therefore fixed within a distance of about 30 m. From another point of view, the maintenance of such an accuracy implies that the geometric corrections to the arrival times have been made very accurately; for example, this accuracy through a single day's observation is sufficient to measure the Earth's radius within 1 km.

Correction for the rate of change of period is clearly necessary on the time scale of one day. Using a linear change $\dot{P}$ only, the arrival time τ of the pulses becomes after time t

$$\tau = NP + \frac{1}{2}\frac{\dot{P}}{P}t^2. \tag{7.7}$$

7.4 The Crab Pulsar

The accumulated correction $\frac{1}{2}(\dot{P}/P)t^2$ becomes one-tenth of a period in only one day. The precision of the timing observations is such that the 'age' $P/\dot{P}$ of the Crab Pulsar can be measured in one day to an accuracy approaching 1%.

On longer time scales various irregularities of pulse timing become apparent. The most famous is the step in pulse arrival time which occurred on 28 or 29 September 1969, an event which became known as the 'glitch'. This was first reported as discontinuous change of period, amounting to a decrease of about 2 parts in 10^{10} (Boynton, Groth, Partridge & Wilkinson, 1969). It was later realised, however, that the glitch may also be represented as a small phase advance in the pulse arrival times. Then, the event of 28/29 September 1969 is represented well by an advance of 3 ms, i.e. about one-tenth of a period, which returns part way towards zero within about 5 days. Clarification of this behaviour came only after careful comparison with the much larger glitch observed in the Vela Pulsar (see below).

There are now several series of timing observations available for the Crab Pulsar. A combination of optical results from four observatories over the period 10 December 1969 to 13 April 1970 shows both the available accuracy and some of the problems of interpretation (Horowitz *et al.*, 1971). There were initially some cumulative errors between the reduced results from different observatories; these were found to be due to small differences in the planetary ephemerides and in the assumed pulsar positions. When these were removed the results were combined, and a best fit to arrival times was constructed. The expected behaviour of the pulsar can be expressed in terms of pulse frequency ν and its derivatives, so that the expected pulse number (N) at time t should be obtained with sufficient accuracy from the cubic equation

$$N = \nu_0 t + \tfrac{1}{2}\dot{\nu}_0 t^2 + \tfrac{1}{6}\ddot{\nu}_0 t^3. \tag{7.8}$$

For assumed values of parameters ν_0, $\dot{\nu}_0$ and $\ddot{\nu}_0$, the expected values of N are found at the observed times t. Generally N is not then integral; the fitting process is the adjustment of the parameters to minimise the fractional part of N over the whole run of observations. When a best fit has been made the residuals may represent either a genuine irregularity in behaviour of the pulsar, or a failure of the cubic law to represent a regular behaviour. The residuals found in this series of observations are shown in Fig. 7.2 as differences between expected and measured arrival times.

This result shows a periodic term with two cycles covering the 150 days of observation. This is not, however, to be taken as evidence of any special physical phenomenon on this time scale, as will be seen from the

Pulse timing

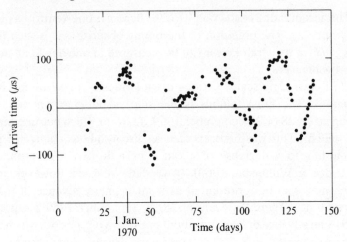

Fig. 7.2. Irregularities in pulse arrival time for the Crab Pulsar. Observations from four optical observatories are combined in this graph. Typical observational errors are less than 10 μs. (After Horowitz *et al.*, 1971.)

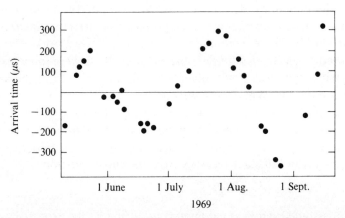

Fig. 7.3. Irregularities in radio pulse arrival time for the Crab Pulsar. Observations from Arecibo Radio Observatory. (After Richards, Pettengill, Counselman & Rankin, 1970.)

results of a different period of observations using radio pulses at 408 MHz (Fig. 7.3). Again a best fit to a cubic curve has been made, and again a periodic term remains; but the period is related again to the length of the series of observations. Evidently the periodic terms are an artifact depending on the assumption of the cubic law in (7.8). A different form of residual is in fact obtained from an only slightly different analysis, in

7.4 The Crab Pulsar

which the arrival times t are related to the pulse number N by

$$t = t_0 + P_0 N + \tfrac{1}{2} P_0 \dot{P} N^2 + \tfrac{1}{6} P_0^2 \ddot{P} N^3. \tag{7.9}$$

There is no *a priori* reason to prefer this expression to (7.8) above. The situation is that the accuracy of the observations has overtaken the precision within which either cubic law can be applied.

One firm conclusion is that the cubic law fits the timing over long periods within about $100\,\mu$s, i.e. within about $10°$ of rotation of the neutron star. We may, however, remark on the short-term fluctuations in pulse frequency in Fig. 7.4, from the radio observations of Roberts &

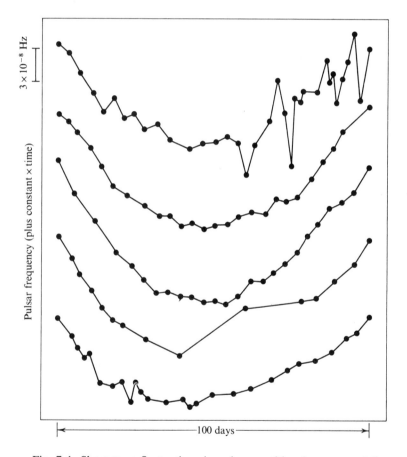

Fig. 7.4. Short-term fluctuations in pulse repetition frequency of the Crab Pulsar. Each span of 100 days should follow a smooth curve whose curvature is determined by the second derivative $\ddot{v}$. The glitch of September 1970 is in the second curve from the bottom. (Arecibo data, after Roberts & Richards, 1971.)

Pulse timing

Richards (1971). These might be made up of small events like the glitch of September 1969. This interpretation has been questioned by Boynton *et al.* (1972) who have attempted to fit the deviations from smooth behaviour by models in which the rotation is modulated in a random way, either as a frequency modulation or as a phase modulation. They found that the best fit was obtained by a random frequency modulation, using a Gaussian modulation function. We discuss this further in connection with the interpretation of the glitches in Section 7.6 below.

7.5 The Vela Pulsar PSR 0833−45

This pulsar has a period of 89 ms; only the Crab Pulsar and PSR 1913+16 have shorter periods. The rate of change of period is easily measurable; it corresponds to the comparatively short lifetime (about 10^4 years) known to be associated with the Vela Nebula. The period increases by about 10.7 ns per day.

Between 24 February and 3 March 1969, the period of this pulsar decreased by about 200 ns, subsequently recovering over several weeks to a period close to, but not identical with, that expected by extrapolation from earlier observations (Fig. 7.5). The 'glitch' observed in PSR 0532+

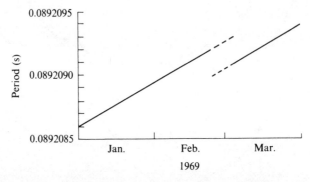

Fig. 7.5. Step in period of the Vela Pulsar PSR 0833−45. At some date between 24 February and 3 March 1969 the period decreased by 200 ns, a change of 2 parts in 10^6. This 'glitch' is superposed on the normal smooth increase of period.

21 may be regarded either as a jump in frequency or a jump in phase; in any case the event constituted a jump in phase of only one-tenth of a rotation, so that there was no difficulty in following the pulse sequence through the glitch without losing count. In the Vela Pulsar the glitch is quite different. Regarded as a phase discontinuity, the glitch introduced an error of some hundreds of rotations. As a frequency discontinuity, it

7.6 Neutron star structure and the glitch function

amounted initially to 2 parts in 10^6, followed by a smooth partial recovery. A description as a frequency discontinuity is clearly more appropriate.

A second such glitch was observed in the Vela Pulsar in 1972. The fact that two such catastrophic events can occur with such a short time interval is of great significance in the interpretation of the internal structure of this pulsar (see Section 7.7).

7.6 Neutron star structure and the glitch function

The discontinuous changes in rotation speed, known as glitches, provide a remarkably penetrating means of investigating the interior structure of the pulsars. The sensitivity of these observations is demonstrated by considering the small change in radius R which, if the neutron star shrank uniformly, would give the observed speed-up of 1 part in 10^9 (corresponding to the Crab glitch of September 1969). The corresponding change in moment of inertia I is given by

$$\frac{\Delta I}{I} = \frac{2\Delta R}{R} = -\frac{\Delta \Omega}{\Omega}. \tag{7.10}$$

For a radius of 10 km the change would be only 5 μm, a very remarkably small quantity to be observed at a distance of 2 kpc.

The glitch observed in the Vela Pulsar was much larger than that observed in the Crab Pulsar, and its form is known in more detail. Taking account of a steady value of $\dot{P}$, the period changes took the form of Fig. 7.6, showing an initial step and a exponential recovery to an intermediate value.

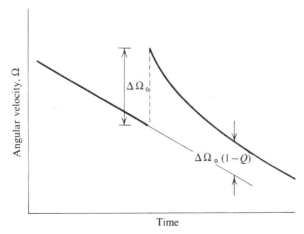

Fig. 7.6. The 'glitch function'. The steady decrease in angular velocity is interrupted by the step Ω_0, which is followed by a partial recovery.

Pulse timing

This behaviour is well represented by the expression

$$\Omega(t) = \Omega_0(t) + \Delta\Omega_0[Q\exp(-t/\tau) + (1-Q)] \tag{7.11}$$

which is now known as the 'glitch function' (Baym, Pethick, Pines & Ruderman, 1969b). The glitch observed in the Crab Pulsar can also be fitted by this expression. The best fit values for the two pulsars are given in Table 7.1 (see, for example, Pines, 1974).

TABLE 7.1 *Observed parameters of the glitch function*

	$(\Delta\Omega)_0/\Omega$	Q	τ
Crab Pulsar	$8.8 \pm 3.8 \times 10^{-9}$	0.916 ± 0.070	4.8 ± 2.0 days
Vela Pulsar	2.3×10^{-6}	0.145	1.2 years

There has not been another clear-cut glitch in the Crab Pulsar, but the Vela Pulsar underwent a very similar event in August 1971, only 17 months after the first event. The six entries in Table 7.1, together with the Vela repetition, form a remarkably useful basis for interpretation.

The form of the glitch function shows immediately that the pulsar does not rotate as a single rigid body, but that it has two components which are only loosely coupled together. One component, which for the Crab Pulsar is solid crust, has a moment of inertia I_c. This component undergoes a sudden change of inertia ΔI, and changes its angular velocity by $\Delta\Omega_0$ where

$$\frac{\Delta I}{I_c} = -\frac{\Delta\Omega_0}{\Omega_0}. \tag{7.12}$$

The second component, which is the neutron liquid, is not initially affected, and continues with the same angular velocity. The liquid and solid components then exert a frictional couple on each other, and come slowly into co-rotation, with a time constant τ. The final change in angular velocity is $(1-Q)\Delta\Omega_0$, which is related to the total moment of inertia I_t by

$$\frac{\Delta I}{I_t} = -\frac{\Delta\Omega_0}{\Omega_0}(1-Q). \tag{7.13}$$

Hence the factor Q is the ratio of the liquid to the solid components of I.

It is immediately clear from the long delay τ that the frictional forces are extremely small, which indicates that the neutron fluid is in a superfluid state. All the solid parts of the neutron star, i.e. the crust and any solid core, rotate as a unit, because they are tightly coupled magnetically. The solid parts are also tightly coupled to the electron and proton components of the fluid. The differential rotation means that the electrons and protons move freely through the neutron fluid, a situation which itself is only possible for more than a fraction of a second if the proton component is itself superfluid (Baym *et al.*, 1969a, b). The small frictional force corresponding to the observed relaxation time τ is between the electrons and a very small proportion of the neutrons which are normal and not superfluid. This small fraction is located inside the cores of vortex lines which form in the rotating superfluid.

The second deduction is made from the values of Q. For the Crab Pulsar the value $Q \approx 0.9$ is entirely consistent with a model neutron star with mass approximately $0.5 M_\odot$. The much lower proportion of neutron liquid in the Vela Pulsar, where $Q \approx 0.15$, allows of two interpretations. The neutron star mass may be low, approximately $0.15 M_\odot$, so that the solid crust extends far into the interior; alternatively the star may be more massive than the Crab Pulsar, with a central density sufficiently high for a solid core to form. As will be seen in the next section, the choice between these can be made through consideration of the magnitude and frequency of the glitches.

7.7 Starquakes and corequakes

A change in inertia ΔI is to be expected from time to time as the rotation of a neutron star slows down. Under centrifugal force, the crust tends to take up an equilibrium oblate form, which becomes progressively less eccentric. The moment of inertia in the equilibrium form is $I_0(1+\varepsilon)$ where the eccentricity ε is determined by the ratio of angular kinetic energy to gravitational energy:

$$\varepsilon = I\Omega^2 \left(\frac{GM^2}{R}\right)^{-1}. \tag{7.14}$$

The strain energy at a different eccentricity $\varepsilon + \Delta \varepsilon$ is proportional to $\mu V(\Delta \varepsilon)^2$, where μ is the shear modulus of the crust, and V its volume.

In the Crab Pulsar the equilibrium value of ε is approximately 10^{-4}; the change of ε in the glitch was therefore approximately 1 part in 10^5. Pines (1974) uses these values to calculate that the stored gravitational energy in the eccentric crust is 2×10^{42} erg, and that the glitch released about 4×10^{39} erg. A glitch of the observed magnitude could therefore be expected to occur every few years during a 'lifetime' of 10^3 years.

For the Vela Pulsar the results of this calculation are totally different. If the much larger glitch is still to be such a 'crust quake', the eccentricity

Pulse timing

would have to change by over 3% in a single glitch, while the energy released would be so great that such events would necessarily be separated by millions of years instead of the observed 2 years. The alternative hypothesis is that the 'quake' occurred in a solid core rather than in the crust (Pines, Shaham & Ruderman, 1972).

The large release of energy, and the short interval between glitches, implies a large value for the eccentricity of whichever part of the star it is that changes shape. This is only possible if the solid part of the star is rigid enough to depart considerably from the equilibrium eccentricity; this can occur only for core material, in which abnormally high values of rigidity are to be expected. The stored energy for this core model is 4×10^{47} erg, and the release in the glitch was 8×10^{44} erg. The corequakes of the Vela Pulsar are expected to continue over a period of some thousands of years, each quake representing a small re-adjustment of the core towards an equilibrium spherical shape.

In summary, one may say that the glitches have provided remarkable confirmation of the calculated models of neutron star structure, and have shown that the Vela Pulsar has a substantial core, the existence of which was only partially expected from theory. Furthermore, the small random variations of frequency in the Crab Pulsar (the 'restless' behaviour), are now simply explained as small starquakes, minor local adjustments of the crustal material, occurring between the cataclysmic re-adjustments of oblateness in the glitches.

The majority of the pulsars seem to lead a comparatively calm life, without appreciable starquakes. Apart from the two 'fast' pulsars, only one has shown a discontinuity in period. Manchester & Taylor (1974) found a small step in period of PSR 1508+55, amounting to a fractional decrease of 2.2×10^{-10}. If such steps were common, they would eventually show up in the annual measurements of period changes which have been made for several years at Jodrell Bank (Lyne *et al.*, 1975). This series of measurements shows no instance of a *decrease* of period as measured over a complete year, and any discontinuities would only show as irregularities in the observed annual increase in period.

7.8 The age of pulsars and the braking index

The slowdown rate $\dot{P}$ has been measured for the majority of the pulsars. The ratio $P/\dot{P}$, which has the dimension of time, is often referred to as a pulsar age; for the Crab and Vela Pulsars the determination of this age was a key step in their identification with supernova remnants. We shall see later (Chapter 20) that this may not be the true age for the older pulsars, so that it should properly be referred to as the 'present time scale

7.8 Age of pulsars and the braking index

of evolution'. Nevertheless, we assume first that neutron stars are formed with rotation periods of only a few milliseconds, and subsequently evolve according to a simple law, which for example might follow the energy loss from a rotating magnetic dipole.

It is convenient to assume that the slowdown law is a power law in angular velocity Ω, of the form

$$\dot{\Omega} = -k\Omega^n \tag{7.15}$$

where k is a constant and the index n is known as the 'braking index'. When the energy loss dw/dt is simply due to the electromagnetic radiation from a rotating dipole, moment m, then

$$\frac{dw}{dt} = I\Omega\dot{\Omega} = -\tfrac{2}{3}m^2\Omega^4 c^{-3} \tag{7.16}$$

and the braking index $n = 3$.

The braking index can be measured directly through direct measurements of the second differential of Ω or P, which can only be made so far for the Crab Pulsar. In terms of angular velocity Ω or frequency ν:

$$n = \frac{\Omega\ddot{\Omega}}{\dot{\Omega}^2} = \frac{\nu\ddot{\nu}}{\dot{\nu}^2} \tag{7.17}$$

while in terms of period

$$n - 2 = \frac{P\ddot{P}}{\dot{P}^2}. \tag{7.18}$$

Experimental values for the Crab Pulsar suggest that n is not constant, but varies between about 2.5 and 3.0, with a long-term mean of about 2.6 (Roberts & Richards, 1971). Although this clearly departs from the simple electromagnetic interpretation, implying a lower rate of energy loss, the difference may easily be accounted for in terms of magnetospheric models in which the field pattern is modified by the stellar wind (Mestel, 1971).

If we assume that n is approximately 3 for other pulsars, and that the slowdown law does not change through the pulsar's lifetime, we can assign an age to any pulsar from a measurement of P and $\dot{P}$. Integration of (7.15) leads to an expression for the time t of pulse number N:

$$N = N_0 + a\left(1 + \frac{t}{T}\right)\frac{1}{n-1} \tag{7.19}$$

where a, N_0 are constants and T is a lifetime. The measured ratio $P/\dot{P}$ is

Pulse timing

related to this lifetime by:

$$\frac{P}{\dot{P}} = T\frac{n-1}{n-2} \tag{7.20}$$

so that for $n = 3$ the lifetime $T = \frac{1}{2}P/\dot{P}$. Some authors have for this reason used $\frac{1}{2}P/\dot{P}$ as an 'observed' value for pulsar lifetime.

The doubts about the validity of this method of measuring pulsar lifetimes, which are set out in Chapter 20, are concerned with the assumption that k in (7.15) is in fact constant. It is suggested instead that it falls, possibly due to an exponential decay in B_0; the result is that $\frac{1}{2}P/\dot{P}$ represents a present time scale of evolution which is much longer than the actual age of the pulsar.

7.9 Pulsar proper motion

The velocities of pulsars are believed to be much higher than those for most stars, and it is reasonable to suppose that the proper motions of many pulsars may be measurable. There is direct optical evidence that the Crab Pulsar is no longer at the centre of expansion of the surrounding nebula, giving a proper motion of about $0''.01$ per year (Trimble, 1971). Proper motions have been measured for a few pulsars by an interferometric method, and for one pulsar (PSR 1133+16) by means of timing observations made over a 4-year period (Manchester *et al.*, 1974).

The proper motion of PSR 1133+16 was detected as a progressively increasing error in pulse arrival times. At the beginning of the 4-year period a best-fit pulsar position was obtained by the method of Section 7.1 above. The difference between predicted and observed pulse arrival times then built up as shown in Fig. 7.7. The linearly increasing sinusoid corresponds to a proper motion, whose amplitude and phase depend on the direction of the motion in ecliptic co-ordinates. The observed proper motion components in right ascension μ_α and declination μ_δ were:

$$\mu_\alpha = 0''.04 \pm 0''.12 \; yr^{-1}, \qquad \mu_\delta = 0''.56 \pm 0''.24 \; yr^{-1}.$$

The significance of this large proper motion is discussed in Chapter 19.

A large transverse component of velocity gives rise to an appreciable increase in period ($\dot{P}$) even if the pulsar is not actually slowing down. Let the transverse velocity be V, and the distance r. Then the apparent change is

$$\dot{P} = \frac{V^2}{rc} P. \tag{7.21}$$

As a measure of the possible importance of this effect, we can estimate an 'age' $P/\dot{P}$ and compare it with observed values. For example, a pulsar

7.10 Binary systems

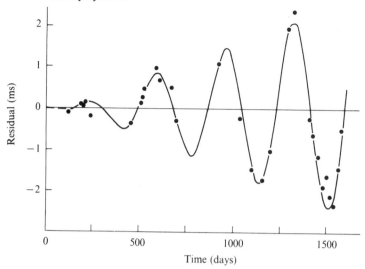

Fig. 7.7. Proper motion of PSR 1133+16. The growing sinusoidal pattern of errors in pulse arrival time is due to an angular motion of about $\frac{1}{2}$ arc second per year. (After Manchester *et al.*, 1974.)

at the comparatively small distance of 50 pc, with a velocity of 200 km s^{-1} would have an apparent age of 10^8 years. For most pulsars the measured values of $P/\dot{P}$ are of the order of 10^6 to 10^7 years, so that the effect is usually unimportant. For PSR 1133+16 it amounts to 5% of the measured value of $P/\dot{P}$.

7.10 Binary systems

The pulse arrival times for almost all pulsars are notably free from the modulation which is typically found in the X-ray pulsars, and which is due to orbital motion in a binary system. We first establish the limits within which it can be said that the majority of pulsars do not have orbiting companions, which might be planets or companion stars.

The detection of a planet orbiting a pulsar would be difficult. They may, nevertheless, exist; despite the possibility that in a supernova explosion a planetary orbit would be so perturbed that the planet might leave the system. Rees & Trimble (1971) have shown that in the Solar System only Mercury and the asteroids would be lost in a solar supernova explosion. A remaining heavy planet such as Jupiter would displace the barycentre of the remaining system, resulting in an appreciable periodic modulation of pulse arrival time, which has not so far been seen.

A planet with mass m, at mean orbital distance $\bar{r}$, will displace a star with mass M from the barycentre by a distance $(m/M)\bar{r}$. The pulse arrival

Pulse timing

time will then vary periodically with amplitude $m\bar{r}/Mc$. For example, the displacement of the centre of the Sun from the barycentre by Jupiter corresponds to 2.5 seconds. The orbital period of Jupiter is 11.9 years; in 1 year the timing of a pulsar with this displacement could change by about 800 ms.

Not many pulsars have been timed over a long enough period to reveal such a modulation, if it does exist. A continuation over several years of the measurements intended to provide values of $\dot{P}$ should at least provide strict limits on the possibility of heavy planets. Lighter planets, such as the Earth or Mars, would produce a modulation only of the order of 1 ms; there is little hope of observing such a small effect except in short-period pulsars such as the Crab Pulsar.

Binary star systems are commonly found, and a considerable proportion of the neutron star population might derive from binary systems in which a supernova explosion has not led to complete disruption (Chapter 12). The period P_b (in years) of a binary star system, with star masses M_1 and M_2 (in solar units), and with semi-major axis a measured in astronomical units, is given by

$$P_b^{-2} a^3 = M_1 + M_2. \tag{7.22}$$

The peak-to-peak variation of pulse arrival time, assuming the major axis is aligned along the line of sight, is $1000a$ s. If timing observations are made throughout the binary period, a sinusoidal variation of pulse delay will be observed, with amplitude $500a \sin i$ s, where i is the inclination of the pole of the orbit to the line of sight. Typically, this would easily be observed if it amounted to a few milliseconds, particularly if the orbital period were as short as a few days. For a typical X-ray binary the orbital period is only a few hours; for $M_1 + M_2 = 1$ the peak delay is then of order 10 s, which would be noticed immediately.

An orbital period very much longer than the total range of timing observations might not be correctly interpreted, since it would appear only as a linear variation of arrival time, i.e. as an additive term of $\dot{P}$ which can be either positive or negative. The observed values of $\dot{P}$ are all positive, so that there can be no large effects from binary orbital motion. In terms of age $P/\dot{P}$, the effect generally cannot exceed 10^6 years, corresponding to a change of 1 part in 10^6 over 1 year. It is a complicated matter to place precise limits on binary systems which could exist within this limit, but an analogy with the terrestrial orbit round the Sun provides an illustration. The maximum orbital velocity of the Earth is 30 km s^{-1}, so that a pulsar with low mass, orbiting as a companion to the Sun, at distance a and binary period P_b, would have a velocity of $30a/P_b$ km s^{-1}.

7.11 *The binary pulsar PSR 1913+16*

The greatest velocity difference in 1 year, assuming $P_b \gg 1$ year, is then

$$\frac{30}{2\pi}\frac{a}{P_b^2}\text{km s}^{-1}$$

which is a fraction $1.5\times 10^{-5}a/P_b^2$ of the velocity of light. This fraction must be less than 10^{-6} to escape detection. Evidently a long-period binary, say with $P_b > 10$ years, might easily have remained undetected.

We conclude that the lack of observed modulation in most pulsars still allows the existence of comparatively low mass planets, and comparatively long-period binary companions.

7.11 The binary pulsar PSR 1913+16

Well over a hundred pulsars had been discovered before one was found to be a member of a binary system. The pulsar PSR 1913+16 was discovered by Hulse & Taylor (1974) during a systematic search using the Arecibo radio telescope (Chapter 2). The orbital period of $7\frac{3}{4}$ hours is remarkably short: so short, in fact, that the rapid changes of period due to changing Doppler shift made its detection and confirmation particularly difficult. The modulation of pulse arrival time is shown in Fig. 7.8. No eclipsing occurs.

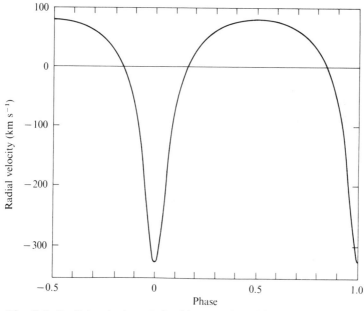

Fig. 7.8. Radial velocity of the binary pulsar PSR 1913+16. The velocity is found from the modulation of the pulse period due to the Doppler effect. The curve is markedly non-sinusoidal, indicating the large eccentricity of the orbit. (After Hulse & Taylor, 1974.)

Pulse timing

The curve in Fig. 7.8 is fitted to the modulation expected from a pure elliptical orbit. All experimental points fit this curve with remarkable precision. The main characteristics of the pulsar and its orbit are as follows:

Period	0.0590301 s
Orbital period	27906.95 s
Semi-amplitude of velocity curve	199.9 km s^{-1}
Eccentricity of orbit	0.6144 ($\pm$0.0010)
Longitude of periastron	178.7 deg
Advance of periastron	3.6 $\pm$ 1.6 deg yr^{-1}
Dispersion measure (*DM*)	167 $\pm$ 5 pc cm^{-3}

The timing observations are capable of great precision, and it should be possible to observe fairly soon some quite small changes in the period and orbital characteristics. The advance of periastron is of particular interest, because in such a small and highly eccentric orbit it is very sensitive to the nature of the companion star. If this is a main sequence star, the orbit is very close to the surface at periastron, and a large apsidal motion (about 5° per orbit) would be expected. This possibility is already excluded. The companion may be a more compact massive star, such as a helium star (which may be observable optically), or it may be a condensed star such as a white dwarf or neutron star. If it is a condensed star, the apsidal motion is due to general relativity only, and the possibility arises of determining separately the two individual masses of the binary system.

This binary system may turn out to be a textbook example in relativistic dynamics. Hulse & Taylor point out that the changes in gravitational potential through the elliptic orbit, and the changes in the relativistic factor $(1-v^2/c^2)^{-1/2}$ through velocity changes in the elliptic orbit, both affect the orbit by measurable amounts. The effects may then replace such phenomena as the precession of the orbit of Mercury as tests of relativistic theory.

References

Baym, G., Pethick, C. & Pines, D. (1969a). *Nature, Lond.* **224**, 673.
Baym, G., Pethick, C., Pines, D. & Ruderman, M. (1969b). *Nature, Lond.* **224**, 872.
Boynton, P. E., Groth, E. J., Hutchinson, D. P., Nanos, G. P., Partridge, R. B. & Wilkinson, D. T. (1972). *Astrophys. J.* **175**, 217.
Boynton, P. E., Groth, E. J., Partridge, R. B. & Wilkinson, D. T. (1969). *IAU Circular No. 2179*.
Horowitz, P., Papaliolios, C., Carleton, N. P. *et al.* (1971). *Astrophys. J.* **166**, L91.
Hulse, R. A. & Taylor, J. H. (1974). *Astrophys. J.* **195**, L51.

References

Hunt, G. C. (1971). *Mon. Not. R. astron. Soc.* **153**, 119.
Lyne, A. G., Ritchings, T. & Smith, F. G. (1975). *Mon. Not. R. astron. Soc.* **171**, 579.
Manchester, R. N. & Taylor, J. H. (1974). *Astrophys. J.* **191**, L63.
Manchester, R. N., Taylor, J. H. & Van, Y. Y. (1974). *Astrophys. J.* **189**, L119.
Mestel, L. (1971). *Nature Phys. Sci.* **233**, 149.
Pfleiderer, J. (1971). *Astron. Astrophys.* **13**, 496.
Pines, D. (1974). *Proc. 16th Solvay Conf. on Physics, Brussels.* p. 147.
Pines, D. & Ruderman, M. (1969). *Nature, Lond.* **224**, 872.
Pines, D., Shaham, J. & Ruderman, M. (1972). *Nature Phys. Sci.* **237**, 83.
Rees, M. J. & Trimble, V. (1971). *Nature, Lond.* **229**, 395.
Richards, D. W., Pettengill, G. H., Counselman, C. C. III & Rankin, J. M. (1970). *Astrophys. J.* **160**, L1.
Roberts, J. A. & Richards, D. W. (1971). *Nature Phys. Sci.* **231**, 25.
Trimble, V. (1971). *IAU Symposium No. 46*, p. 3 (Dordrecht: D. Reidel.)

8
Properties of the integrated radio pulses

The most striking characteristic of pulsars is undoubtedly the astonishing regularity of the pulsation period. This machine-like precision may suggest that the whole range of phenomena to be observed in pulsar radiation should follow equally simple patterns; on the contrary, there is an almost bewildering range of variability in the characteristics of the pulses, which sometimes approaches chaos. The actual time of arrival of the individual pulses varies over a considerable range, their strength varies on several distinct time scales, and their polarisation is variable. The task of the observer is to describe this complex situation as simply as possible.

Useful descriptions can be made either by isolating typical properties of individual pulses, as for example their average width, or alternatively by superposing many pulses and describing an integrated pulse profile. Many pulsars are in practice too weak for individual pulses to be detected, so that only the integrated pulse profile can be studied. This, however, does prove to be characteristic of an individual pulsar, and the duration and shape are often, though not always, independent of the radio frequency. Details of the integrated profiles are often used in fitting observations to theories of emission, although we shall see that the interpretations must also take account of the individual pulses. This chapter is concerned only with the integrated profiles; the variations they conceal are dealt with separately in the next chapter.

8.1 Integrated pulse profiles

The integrated pulse profiles are obtained by superposing a sequence of some hundreds of individual pulses. This is achieved by sampling the radio signal at small time intervals, and superposing or 'folding' the sequence of samples at the period of the pulsar. Since the individual pulses are often highly polarised, it is necessary to ensure that the total intensity is recorded; consequently two orthogonal modes of polarisation must be received, separately detected, and added. The signal-to-noise ratio in the integrated profile improves with larger receiver bandwidths and integration times; it is, however, necessary to restrict the bandwidth

8.1 Integrated pulse profiles

for pulsars with large dispersion measures, since there would otherwise be a smearing effect which would spoil the time resolution (Chapter 2).

The integrated profiles for some of the strongest pulsars, recorded at 408 MHz, are shown in Fig. 8.1. The profiles are smooth curves, generally with up to three components, more rarely with four or five. They generally occupy between 2% and 10% of the period. We may ask first whether this fraction is independent of the period, so that the profile length may be expressed in terms of an angular width corresponding to a rotation angle of the pulsar. This is shown to be the case in Fig. 8.2; here the width is expressed as an equivalent width by dividing the area of the profile by its height. The straight line in Fig. 8.2 corresponds to an angular width of 9°. Fig. 8.3 shows the histogram of the equivalent pulse width in degrees of rotation: with few exceptions the widths lie between 3° and 30°. Fig. 8.4 shows the integrated profiles of a larger number of pulsars, plotted on a constant scale of angular rotation.

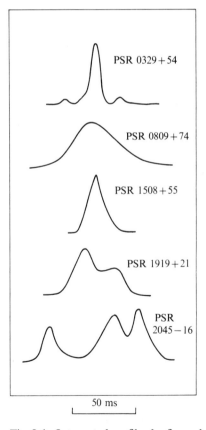

Fig. 8.1. Integrated profiles for five pulsars. (Jodrell Bank; 408 MHz.)

Properties of integrated radio pulses

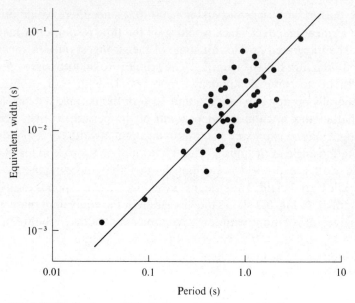

Fig. 8.2. Equivalent width versus period. The straight line represents an angular width of 9°.

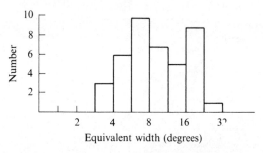

Fig. 8.3. Histogram of equivalent width in degrees of longitude.

There is a tendency towards symmetry in all the integrated profiles, which are generally of the following types:

(i) a smooth single hump, e.g. PSR 1642−03;
(ii) a double hump, e.g. PSR 1133+16;
(iii) a single hump with extensions or 'outriders', e.g. PSR 0329+54;
(iv) a double hump with structure between, e.g. PSR 1237+25.

These profiles will be interpreted later (Chapter 18) as distributions of emitting regions over a range of longitude in the pulsar magnetosphere.

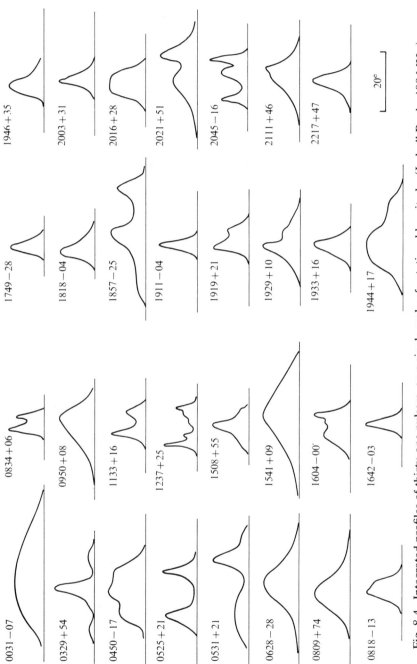

Fig. 8.4. Integrated profiles of thirty-one pulsars, on a single scale of rotational longitude. (Jodrell Bank; 408 MHz.)

Properties of integrated radio pulses

8.2 On/off ratio: interpulses

Between the pulses there is a remarkably low intensity, usually below the detection level. An example is shown in Fig. 8.5, which shows the whole period of PSR 1133+16 using an integration time of 2 hours. The ratio of on-pulse to off-pulse intensity is greater than 1000:1.

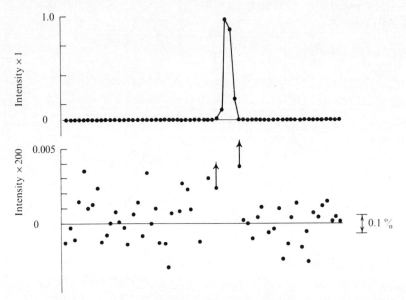

Fig. 8.5. On/off ratio for PSR 1133+16. The top trace shows the whole period (1.18 s) with a resolution of 18 ms. The low level outside the pulse is shown multiplied by 200 in the lower trace. Only random noise can be seen. (Jodrell Bank; 408 MHz.)

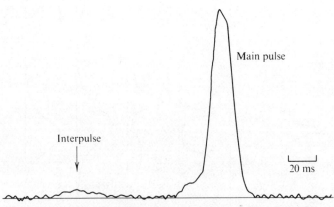

Fig. 8.6. Interpulse in PSR 0950+08. The trace shows the full period (0.253 s). (After Rickett & Lyne, 1968.)

8.3 Frequency dependence of the integrated pulse profiles

For some pulsars a second component, known as an 'interpulse', appears somewhere near, but not exactly at, the halfway point between the main pulses. This is seen for example in PSR 0950+08 (Fig. 8.6), where the interpulse intensity is about 2% of the main pulse intensity (Rickett & Lyne, 1968). The best example of an interpulse is provided by the Crab Pulsar (Chapter 11).

8.3 Frequency dependence of the integrated pulse profiles

Although it is a useful generalisation that the properties of pulsars do not depend markedly on radio frequency (or indeed, for the Crab Pulsar, on the frequency at any part of the electromagnetic spectrum), there are some systematic changes to be found in a number of pulsars. Fig. 8.7 shows the changes in integrated profiles over a 4:1 range of radio frequency. For pulsars with a double profile Craft & Comella (1968) found that the separation between the two peaks tended to increase at lower frequencies, and it is often the case that the separation between identifiable components varies in this way. For PSR 2045−16, which has three large components, the leading and trailing components follow the same rule, moving inwards towards the centre one at increasing frequencies. The amplitude of the trailing component also increases notably at higher frequencies.

PSR 0950+08 and PSR 2016+28 both show very marked changes in profile with frequency. At low frequencies, the component at the front of

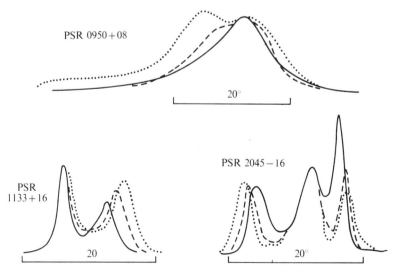

Fig. 8.7. Profiles at three frequencies: ——, 610 MHz; − − − −, 240 MHz; · · · · ·, 150 MHz. (After Lyne, Smith & Graham, 1971.)

Properties of integrated radio pulses

the main pulse from PSR 0950+08 becomes very strong, while the trailing half of PSR 2016+28 almost disappears. The equivalent width of PSR 2016+28 decreases by nearly a factor of two between 408 MHz and 151 MHz. Although there have been attempts to fit these changes to a simple law relating component separation to radio frequency, there is not much evidence that any regular pattern can be distinguished apart from a tendency towards increasing equivalent width at lower frequencies. Obviously the question of regular behaviour of this sort is important in the theory of radiation beaming; the changes in profile might mean changes in the direction of a radiation beam, or they might mean that radiation is emitted with a different spectrum from different points on a distributed source.

8.4 Integrated polarisation

Pulsar radiation is remarkably highly polarised. As we will see in the next chapter, individual pulses are typically 100% elliptically polarised, including cases where full linear or full circular polarisation can be observed. A statistical average of polarisation, made in a similar way to an integrated pulse profile, is less fully polarised, but it takes a precise form for each pulsar. As with the shape of the integrated intensity profile, the changes of polarisation through the profile can be related to models of the emitting regions, with the additional parameter of a vector which characterises the polarised component of the radiation. This vector is probably related to, and may possibly be identical with, the magnetic field direction at the emitting region.

The full specification of arbitrarily polarised radiation requires four parameters. It is customary and convenient to use the four Stokes parameters I, V, Q, U, which may loosely be described respectively as the total intensity, the circularly polarised component, and the two orthogonal components of the linear polarisation (see, for example, Born & Wolf, 1965). These four quantities may be separately integrated for a sequence of pulses. The result may be presented as separate profiles of the ensemble averages; usually the linear components are combined to give an effective linear component $P = (Q^2 + U^2)^{1/2}$ and a position angle $\frac{1}{2}\tan^{-1} U/Q$. (It is important to note that U and Q here are combined after integration, as ensemble averages, although it has not been considered necessary to distinguish these averages by a separate symbol in this text.)

Figure 8.8 shows the integrated polarisation characteristics obtained through this averaging process for PSR 0329+54 at three different radio frequencies. The profiles are very similar, apart from the differing posi-

8.4 Integrated polarisation

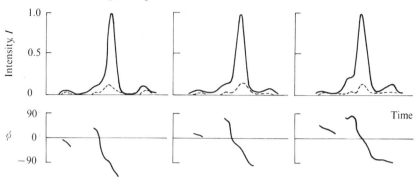

Fig. 8.8. Polarisation of PSR 0329+54 at 240, 408 and 610 MHz from left to right). The upper full trace is the intensity I, the broken trace is the linearly polarised component, and the lower trace is the position angle ϕ.

tions of the 'outrider' components at the three frequencies. There is a smooth swing of position angle across the profile, which continues across the bridge joining the main to the trailing outrider component. The swing of position angle is not significantly different at different frequencies.

Integrated polarisation profiles are available for upwards of thirty pulsars (Lyne, Smith & Graham, 1971; Manchester, 1971; Moffett, 1971). Some of these are shown in Fig. 8.9, including examples of pulsars with complex profiles, such as PSR 2045+16. For some, such as PSR 1911−04, the integrated polarisation is less than 10%; it will be interesting to discuss in the next chapter whether this is a result of averaging over highly polarised but variable individual pulses, or whether the individuals are themselves not highly polarised. Others, such as PSR 1929+10, show very high integrated linear polarisation, reaching at least 90% over much of the pulse profile.

Circular polarisation (Stokes parameter V) appears in many of the integrated profiles, often showing fine structure. In PSR 1237+25, for example, there is a circular component near the centre of the profile which reverses hand so rapidly that the transition is hard to resolve experimentally. The percentage circular polarisation reaches 20% in several pulsars.

The swing in position angle through the integrated pulse profile is characteristically monotonic, covering a range of up to 180° in a single smooth sweep. (Discontinuous steps in position angle do occur in some pulsars; however the few apparent discontinuities in Fig. 8.9 may not be real, being accountable for by a poor signal-to-noise ratio in the observations.) Following the view that the integrated pulse profile represents a

Properties of integrated radio pulses

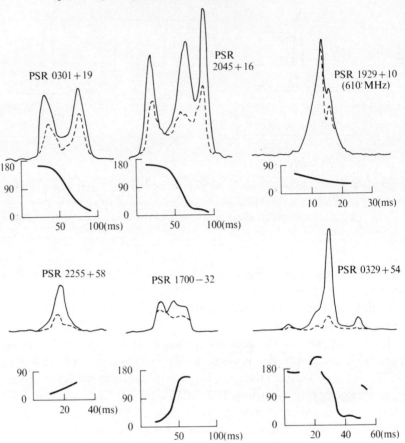

Fig. 8.9. Polarisation in integrated profiles of six pulsars. The broken line represents the linearly polarised component, and the graphs below the profiles show the position angle. (Jodrell Bank recordings at 408 MHz and 610 MHz.)

distribution of emitters over a longitude range in the pulsar magnetosphere, we may interpret this smooth sweep as a range of directions of magnetic field; it then becomes interesting to measure the rate of swing of polarisation position angle ϕ with rotation longitude l (also referred to as 'pulse phase'). Lyne *et al.* (1971) found the modulus of this 'phase rate' generally lay in the range 0 to 15. Five pulsars found to have larger phase rates were doubtful cases, since they showed complex profiles and were not highly polarised. The most highly polarised sources have phase rates of 6 or less. This result is important in the geometrical interpretations of Chapters 17 and 18.

8.6 Mode changing

8.5 Fine structure in integrated profiles

The integrated profiles of total intensity I are usually smooth, often showing only one component. The circular component V does, however, show fine structure, as seen for example in PSR 1237+25. Hankins (1973) has shown that the I profile of PSR 1919+21 contains a similar fine structure, in the form of a notch 1.4 ms wide (Fig. 8.10). The whole profile is over 30 ms long. The position of the notch, on the first main peak of the profile, coincides with a rapid change in position angle of polarisation. The notch was observable at 74 and 111 MHz, but not at 318 MHz.

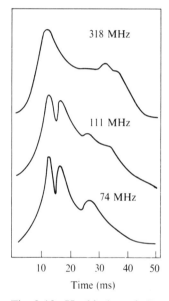

Fig. 8.10. Hankins' notch. Integrated pulse profiles for PSR 1919+21 at three frequencies. (After Hankins, 1973.)

Possibly the sharp features in the pulsars PSR 1237+25 and PSR 1919+21 are similar. In both, the time scale of the fine structure is close to that of the individual pulses, and both coincide with rapid changes in position angle.

8.6 Mode changing

The integrated profiles both of intensity and of polarisation are usually remarkably stable from one sequence of pulses to the next. Some pulsars do, however, show occasional abrupt changes in the profile, maintaining a new configuration for a long sequence of pulses, and returning as abruptly to the original state. This was noted first for PSR 1237+25 by Backer

Properties of integrated radio pulses

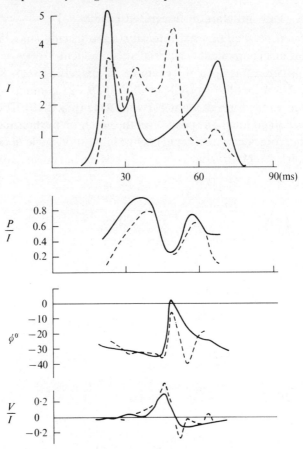

Fig. 8.11. Mode change in PSR 1237+25. The integrated profile (top trace) changes as shown between the full and broken lines. The polarisation characteristics P/I, ϕ, V/I change very little: the differences may be due only to measurement errors.

(1970). Fig. 8.11 shows observations of this pulsar by Lyne (1971), including the polarisation parameters P/I, ϕ, and V/I for both modes. Following Backer, the various peaks in the intensity profile are labelled I to V. In the abnormal mode, the trailing components IV and V are greatly reduced, while the central component III increases relative to component I.

It is particularly important that the polarisation characteristics are hardly changed at all at the mode change. This suggests that the directions of magnetic field associated with the different parts of the profile do not change at the mode change, but rather that there is a change in the relative excitation of different emitting regions.

8.6 Mode changing

Mode changing has not been recognised in many pulsars, but this may only be a consequence of the difficulty of recognition of such changes in weak pulsars. A change in the very strong pulsar PSR 0329+54 was found rather by accident when a long sequence of integrations was made at Jodrell Bank as a demonstration during an IAU Symposium on the Crab Nebula. Over a 6-hour period only one mode change occurred; this took the form of an interchange of intensities between the outrider components (Fig. 8.12a), a comparatively subtle change that could easily be missed in weaker pulsars. This particular mode change is the slowest known so far; typically the pulsar remains in the normal mode for about 4 hours, and in the abnormal mode for about half an hour. There is, however, no regular periodicity in the alternation, and either mode may persist for 8 hours or more.

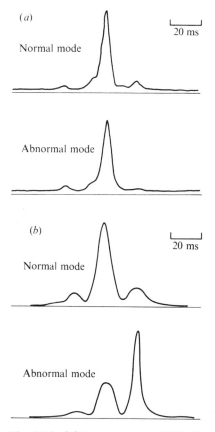

Fig. 8.12. (a) Mode change in PSR 0329+54 at 408 MHz. The change is in the relative intensities of the 'outriders'. (After Lyne, 1971.) (b) Mode change in PSR 0329+54 at 10 GHz. (After Hesse et al., 1973.)

Properties of integrated radio pulses

The mode change in PSR 0329+54 have been shown to occur simultaneously at widely spaced frequencies. The changes are more marked at high frequencies: Fig. 8.12(*b*) shows the two modes at 10 GHz (Hesse, Sieber & Wielebinski, 1973).

8.7 The spectrum of pulsar radio emission

Radio observations of the stronger pulsars extend over about 6 octaves, and several spectra have been determined fairly well over this range. It must again be emphasised, however, that these observations are of the averaged, or integrated, signal; the apparent simplicity and smoothness of the spectra described in this section undoubtedly conceals a more complex behaviour in the individual pulses. Even the integrated spectrum is difficult to obtain with accuracy, because of the inherent variability of the pulsars and the large variability impressed on their radiation by scintillation in the interstellar medium. McLean (1973) has shown that simultaneous observations at different frequencies, using wide bandwidths and long averaging times, can give repeatable spectra, but that for some pulsars the emitted spectrum must be variable.

A collection of all available information by Sieber (1973) contains the spectra shown in Fig. 8.13. All known spectra show a fall of flux density s with frequency ν according to a power law $s \propto \nu^\alpha$, where the index α is -3 ± 1. This smooth fall extends over a wide frequency range for some pulsars, including the Crab Pulsar (see Chapter 11), but for many pulsars the slope is less at low frequencies. For some pulsars, e.g. PSR 0329+54, there is a reversal of slope, giving a maximum flux density at a frequency in the range 100 to 400 MHz. It seems possible that the slope also becomes greater at high frequencies, typically above 1000 MHz, but this is not yet established except for a few pulsars.

8.8 Long-term variations of flux density

Over the whole available range of time scales on which pulsars can be observed, covering from a few microseconds to a few millions of years, there seems to be practically none on which there is no variability of the radio signal. The shortest time scales, from seconds downwards, are concerned with the beaming and radiation processes, while the longest are concerned with the decay of the whole pulsar phenomenon as the neutron star rotation slows down. Between these there is a time scale typically of some minutes, but strongly dependent on frequency, which is associated with interstellar scintillation. The effects of scintillation, and the rapid variations within the source, are easily removed in the averaging process of a typical observation using a wide bandwidth and an integra-

8.8 Long-term variations of flux density

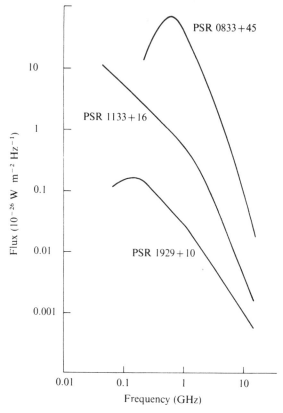

Fig. 8.13. Typical radio frequency spectra of pulsars. The spectra are generally curved, and often show a low frequency cut-off.

tion over many pulses, but large variations still remain with time scales of days and longer. These are referred to as the 'long-term variations'.

Cole, Hesse & Page (1970) showed that for five pulsars observed daily for a year at 81.5 MHz the long-term variations covered a range of over 10:1, with an autocorrelation width typically of 1 month. McLean (1973) showed that these variations showed a definite correlation between frequencies of 150, 240 and 408 MHz. In particular, there seems to be no frequency dependence of the time scale of the variations. No form of scintillation, in which the long-term variations would be due to some property of the transmission path, can account for this frequency-independent behaviour, which must occur within the pulsar.

The origin of these long-term variations remains a complete mystery. The magnetosphere of a rotating neutron star is fairly well isolated from any variable conditions outside the velocity-of-light cylinder, while the models of the neutron star itself, though doubtless much oversimplified, do not seem to allow for slow changes in energy output. So far as known, there are no changes in the integrated pulse profiles between weak and strong periods of emission; this may of course be accounted for partly by observational selection, in which the strong periods are those for which the best profiles were obtained.

References

Backer, D. C. (1970). *Nature, Lond.* **229**, 1297.
Born, M. & Wolf, E. (1965). *Principles of Optics*, 3rd ed. (Oxford: Pergamon Press), p. 554.
Cole, T. W., Hesse, H. K. & Page C. G. (1970). *Nature, Lond.* **221**, 525.
Craft, H. D. & Comella, J. M. (1968). *Nature, Lond.* **220**, 676.
Hankins, T. H. (1973). *Astrophys. J.* **181**, L49.
Hesse, K. H., Sieber, W. & Wielebinski, R. (1973). *Nature Phys. Sci.* **245**, 57.
Lyne, A. G. (1971). *Mon. Not. R. astron. Soc.* **153**, 27P.
Lyne, A. G., Smith, F. G. & Graham, D. A. (1971). *Mon. Not. R. astron. Soc.* **153**, 337.
McLean, A. I. O. (1973). *Mon. Not. R. astron. Soc.* **165**, 133.
Manchester, R. N. (1971). *Astrophys. J. Suppl.* **23**, 283.
Moffett, A. T. (1971). *IAU Symposium No. 46*, p. 195. (Dordrecht: D. Reidel.)
Rickett, B. J. & Lyne, A. G. (1968). *Nature, Lond.* **218**, 934.
Sieber, W. (1973). *Astron. Astrophys.* **28**, 237.

9

Individual radio pulses

The well organised and characteristic behaviour of the integrated pulse profiles becomes the more surprising as one examines in greater detail the complexity and variety of the individual pulses that add to make the integrated profiles. The intensity and shape of the pulses varies from pulse to pulse, and often there is structure within a pulse on a very short time scale. A sequence of a few pulses may present such chaotic variations that it is even hard to believe that the sum of any sequence of only a few hundred pulses can yield the characteristic integrated profile. But there is often a quite simple statistical description of the pulse behaviour, and there are precise rules governing much of the apparent chaos. For example, the energy in a single pulse, as measured at a particular part of the radio spectrum, follows a typical statistical distribution, which is near Poissonian for some pulsars and quite different for others; furthermore this distribution is usually well established from a sequence of only a few hundred pulses.

Individual pulses commonly have a width of only one-tenth or less of the width of the integrated profile; they may appear almost at random at any part (or 'phase') of the profile. Their occurrence is not completely random. Often successive pulses will appear as narrow pulses at nearly the same phase, and in some pulsars a sequence of several pulses will be related in this way. A slow 'drift' of phase is often seen, usually towards the earlier part of the profile. Some pulsars radiate pulses with two or more such narrow components. The consistency of this structure, together with a characteristic polarisation observed within the narrow components, has led to their isolation as a basic component of pulsar radiation. They are now known as 'sub-pulses'.

More rapid fluctuations of intensity occur, usually in the pulsars with shorter periods. These fluctuations are known as the 'microstructure'. They include some remarkably short and intense pulse components in the Crab Pulsar: a single one of these 'giant' pulses may last only $10\,\mu$s, but during that time the intensity can exceed the radio intensity from the whole Crab Nebula. The microstructure does not have a continuity from pulse to pulse.

Individual radio pulses

Each of these characteristic features of individual pulses, i.e. the statistical variations of intensity, the sub-pulse structure, pulse drifting, polarisation, and the microstructure, may vary with radio frequency, and may also be different at different phases of the integrated pulse profile. Any presentation is therefore necessarily a selection from among a very rich field of information, in which it is still very much a matter of opinion which are to be regarded as the phenomena with the most basic significance. We present first a selection of the observational data showing both simple and complex behaviour, and we then concentrate mainly on the sub-pulses.

9.1 The structure of individual pulses

The complexity of the pulses is immediately obvious in the sequences of individual pulses displayed in Figs. 9.1–9.4, which show various aspects of the structure. There are some pulsars, such as PSR 1642−03 (Fig. 9.1), in which the individual pulses are not dissimilar from the integrated profile: there is only a small variation of intensity and a small variation of pulse phase (time of occurrence) between successive pulses. In others, such as PSR 0329+54 (Fig. 9.2), the individual pulses are relatively narrower and there may be more than one component. These pulse components are the 'sub-pulses', which in some pulsars show the phenomenon of 'drifting' to earlier phases in successive pulses.

More rapid fluctuations of intensity can be seen in PSR 0950+08; this is the 'microstructure' which will be described later (Section 9.3). Some pulsars show a combination of these varieties of behaviour, as in PSR 1133+16 (Fig. 9.3). This pulsar has a double-peaked integrated profile, although individual pulses seldom show a double structure: it seems that sub-pulses occur randomly and independently at either of the two preferred locations corresponding to the two peaks.

Although the sub-pulses may well be a basic entity, around which interpretations have naturally centred, they only provide a complete description of the whole radiation from a very few pulsars. In Fig. 9.3 some pulses can be seen in which sub-pulses are superposed on longer pulse components; this may be characteristic of several pulsars. A study of PSR 1919+21 by Cordes (1975) shows that the intensity ratio between the sub-pulse and other components varies through the pulse profile. In Fig. 9.4 a sequence of pulses from PSR 1919+21 recorded simultaneously at 111 MHz and 318 MHz shows that the sub-pulse structure is most prominent in the early part of the integrated profile, and also that it is more prominent at the lower radio frequency. These sub-pulses seem to be cut up by the rapid fluctuations of microstructure, but they neverthe-

9.1 Structure of individual pulses

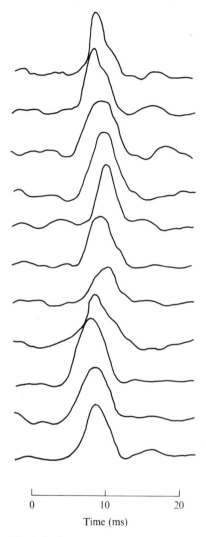

Fig. 9.1. Sequence of pulses from PSR 1642−03 recorded at 408 MHz. (Jodrell Bank recording.)

less show the characteristic drifting. In the later part of the profile, which is more prominent at higher radio frequencies, there are less distinct sub-pulses about 5 ms long.

A most important characteristic of the sub-pulses is their very high degree of polarisation. This was discovered by Clark & Smith (1969) in PSR 0329+54, where the sub-pulses are easily distinguished. The sequence in Fig. 9.5 shows the Stokes parameters for successive pulses from

Individual radio pulses

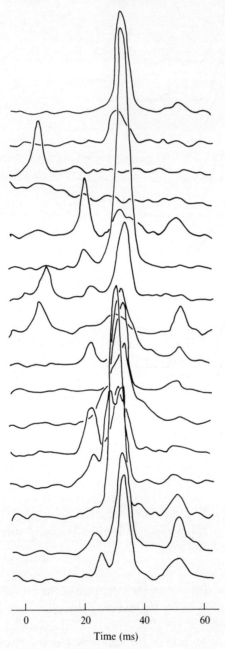

Fig. 9.2. Sequence of pulses from PSR 0329 + 54 recorded at 408 MHz. (Jodrell Bank recording.)

9.1 *Structure of individual pulses*

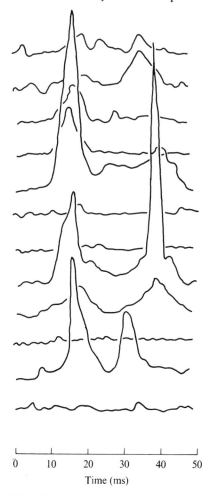

Fig. 9.3. Sequence of pulses from PSR 1133+16 recorded at 408 MHz. (Jodrell Bank recording.)

this pulsar. It is common to find polarisation in excess of 95% in individual sub-pulses. The polarisation is in general elliptical, changing in form through a sub-pulse as, for example, by a smooth change from elliptical through linear to elliptical with the opposite hand. Completely circular, or completely linear, polarisation may occur during a sub-pulse. As far as is known, all clearly defined sub-pulses follow a similar pattern of polarisation, in which there is a smooth, simple sweep of polarisation characteristics through a single sub-pulse. One mode of presentation which shows this sweep is in Fig. 9.6, which displays the circular component as an ordinate and the position angle of the ellipse as abscissa. (A

Individual radio pulses

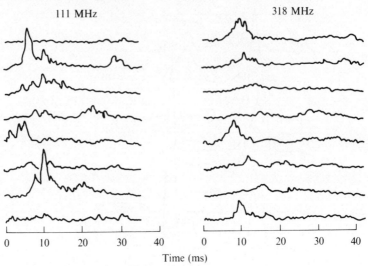

Fig. 9.4. Sequence of pulses from PSR 1919+21 recorded simultaneously at 111 MHz and 318 MHz. A time resolution is achieved by the use of 'de-dispersion'. (After Cordes, 1975.)

plot of this kind is a convenient projection of a Poincaré sphere, on whose surface every point corresponds to a particular state of elliptical polarisation.)

The identification of the sub-pulses as a basic entity depends on their appearance as discrete, symmetrical components, on their coherence and drifting between successive pulses, and on their very high polarisation with its typical swing of characteristics. In contrast, the separate components of an integrated profile do not generally show symmetry or very high polarisation, while on the shorter time scale the microstructure appears as a modulation of intensity in which the polarisation remains unchanged.

9.2 Width of the sub-pulses

The simple symmetrical shape, and the high polarisation, of the sub-pulses suggest strongly that they represent elemental beams of radiation sweeping past the observer as the pulsar rotates. Such a rotating beam is familiar on a terrestrial scale; it can be made by a rotating antenna system, as in an airport control radar. In such an antenna the beamwidth for a given aperture varies inversely as the radio frequency, and by analogy we may expect some such variation in the width of the sub-pulses. It is a very remarkable and important fact that the width and the high polarisation of the sub-pulses are not markedly dependent on frequency (Smith,

9.2 Width of the sub-pulses

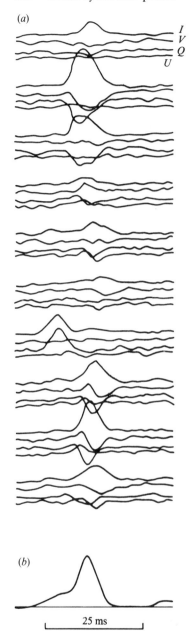

Fig. 9.5. (a) A sequence of pulses from PSR 0329+54 recorded with a polarimeter at 408 MHz. The four traces show the Stokes parameters I, V, Q, U. (b) The integrated profile obtained by addition of the intensity I of several hundred pulses.

Individual radio pulses

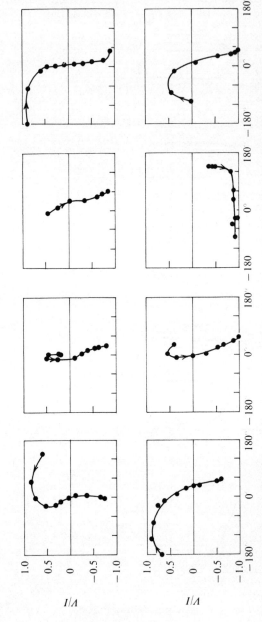

Fig. 9.6. The varying elliptical polarisation of individual pulses from PSR 0329+54. The normalised Stokes parameter V/I is plotted against 2ϕ, twice the position angle of polarisation. The points on the curve are at uniform time intervals of 1 ms.

9.2 Width of the sub-pulses

1970a, b; Manchester, Tademaru, Taylor & Huguenin, 1973; Taylor, Manchester & Huguenin, 1975). This invariance is the main evidence for the relativistic beaming process described in Chapter 17, which is the only known beaming process in which both the width and the full polarisation are independent of radio frequency. The width of a sub-pulse is defined as the full width between half-power points, expressed either as a duration or as degrees of rotation. Values are given in Table 9.1 for those pulsars where sub-pulses have been clearly distinguished.

TABLE 9.1   Width of sub-pulses

PSR	Width ms	Width deg.	Radio frequency MHz	Reference
0031−07	~10	~4	145	2
0329+54	4	2	240, 408, 610	3, 4
		3	150, 410, 1400	7
0628−28	40	11.5	408	3
0809+74	4	1.1	81	1
0823+26	4	2.7	408	3
0834+06	4	1.3	408	3
	~5	1.6	147	4
0943+10	~8	~2.5	147	4
0950+08	1–2	1.5–3	408	3
1133+16	3.5	1.0	408	3
1237+25	6	1.5	408	3
1642−03	4	3.7	408	3
1818−04	~6	~3	400	4
1919+21	6	1.6	408	3
	5	1.3	318	6
2016+28	~5	~3	400	4
2020+28	1.5	1.0	430	5
2045−16	4	0.7	408	3

References:
(1) Cole (1970); (2) Huguenin, Taylor & Troland (1970); (3) Lyne, Smith & Graham (1971); (4) Taylor & Huguenin (1971); (5) Schonhardt & Sieber (1973); (6) Cordes (1975); (7) Manchester et al. (1973).

The angular widths of recognisable sub-pulses lie in the range 1° to 4°, with the exceptions only of PSR 0628−28 and PSR 2045−16. This range is crucial to the interpretation of pulse widths in terms of the relativistic beaming theory (Chapter 17).

Individual radio pulses

9.3 Microstructure

The time resolution available in straightforward observations of radio pulses is usually limited by practical considerations of signal-to-noise ratio. The pulsar PSR 0950+08 is occasionally very powerful, and it also has a very small dispersion measure so that a large signal bandwidth can be used, especially at high radio frequencies. Some exceptionally sensitive observations of this pulsar and PSR 1133+16 were made at low radio frequencies by Hankins (1971, 1972) using a large signal bandwidth with a 'de-dispersion' technique to preserve a high time resolution. The results for PSR 0950+08 are shown in Fig. 9.7.

Recordings of this kind, made with extremely short integration times, must contain some spurious modulation by noise. In the limit, a single radio bandwidth B, observed without smoothing, i.e. with an output time constant less than $1/B$, will display a full Gaussian modulation of any signal. In Fig. 9.7(b) the broken line indicates the expected level of noise modulation for this recording, showing that all the narrow peaks are real.

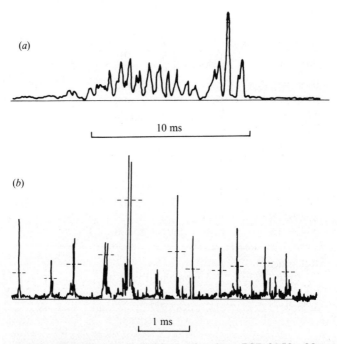

Fig. 9.7. Details of two individual pulses from PSR 0950+08, recorded at 111 MHz using a de-dispersion technique. Time resolution in (a) 112 μs; (b) 7 μs. The broken line in (b) represents a 97.5% confidence limit, showing that the narrow spikes are larger than expected from random noise. (After Hankins, 1971.)

9.4 Histograms of pulse energy

The most intense peaks reach a flux density of 4×10^{-22} W m^{-2} Hz^{-1}; they last for less than 10 μs.

The periodic modulation in Fig. 9.7(a) has only been observed on a few occasions. Although its origin is unknown, it lends support to the view that the microstructure is a modulation of a sub-pulse, rather than a collection of random independent events. Microstructure, with time scales much less than one millisecond, has only been observed in PSR 0950+08. It would have been observed, had it existed, in several other strong low-dispersion pulsars. Possibly its appearance in PSR 0950+08 is related to its other unusual characteristic, namely the apparent confusion of its sub-pulse structure. It is reasonable to maintain that any angular beam pattern of the sub-pulses is swamped by the rapid time variations of the microstructure.

It is a difficult observational problem to measure the bandwidth of the microstructure. In contrast to the wide bandwidth of the sub-pulses, the spectrum of the sharp spikes seems to be narrow. A single sharp pulse, such as one of those in Fig. 9.7(b), lasting a time T must occupy a radio bandwidth of at least $1/T$; for example if $T = 100$ μs the bandwidth will appear to be at least 10 kHz. Whether in fact the spectrum of a single spike is confined to a single such band, or is extended over a much wider range of frequencies, is not yet fully resolved. Rickett has observed some pulses which have a narrow-band structure less than 20 kHz wide, even though the same short pulse is observable over a bandwidth of the order of 100 MHz (Hankins & Rickett, 1975). If this is typical, then it demonstrates that the emission process must be narrow-band, and therefore associated with a resonance at a particular location in the magnetosphere.

9.4 Histograms of pulse energy

The energy of individual pulses from a pulsar evidently varies due to a complex pattern of behaviour within the pulsar. The incidence of sub-pulses within any pulse might be expected to be random, and the size of the sub-pulses might also be expected to follow some form of random statistics, possibly related to a chance occurrence of coherent conditions in the source. We might therefore expect the energy of individual pulses to be distributed according to a recognisable random law.

Histograms of pulse energy from sequences of 1000 pulses recorded at 408 MHz were obtained by Smith (1973), using an analysis which removed the effects of interstellar scintillation. The results for three pulsars are shown in Fig. 9.8.

The histograms are found to repeat remarkably precisely for any individual pulsar. They also differ greatly from one pulsar to another. The

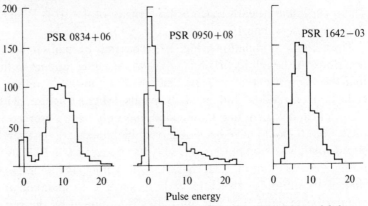

Fig. 9.8. Histograms of pulse energy at 408 MHz. PSR 0834+06 shows a small peak at zero (the missing pulses). PSR 0950+08 shows a maximum at zero, with a monotonic fall. PSR 1642−03 has no missing pulses, and shows a smooth distribution about a single peak.

repetition is shown here for PSR 1642−03, which has a narrow distribution of pulse energy (this distribution as shown in the figure may even be slightly broadened by residual effects of scintillation). The histogram for PSR 0950+08 is entirely different, showing a maximum probability near zero energy, and a monotonic decrease to a long tail of comparatively rare large pulses. The histogram for PSR 0834+06 shows a small peak at zero energy, corresponding to infrequent missing pulses: the width of this peak is due solely to receiver noise. The shape is otherwise intermediate between the other two histograms.

PSR 0950+08 is noted for the complexity of its behaviour; individual pulses occur over a wide range of times within the 'window' defined by the integrated pulse profile, and the integrated profile is different for different radio frequencies. The behaviour of PSR 1642−03 is simple; individual pulses are nearly the same width as the integrated profile (Lyne *et al.*, 1971). Of these two, the comparatively constant performance of PSR 1642−03 is the more remarkable. We will see in Chapter 15 that the radiation process must involve a very high degree of coherence, so that small variations of charged particle density would be expected to produce large variations of emissivity. The constancy of the emission, and the precision with which the histogram of variations repeats in different sequences of pulses, suggests that the radiation process is remarkably well organised.

The greater degree of randomness in PSR 0950+08 may be described in many different ways. It can be fitted to a Poisson distribution in which the probability of amplitude x is proportional to $a^x/x!$; in this case the parameter a is close to 1, while a similar fit for PSR 1642−03 would

9.5 Pulse drifting

require $a = 8$. These descriptions might apply to an ensemble of sources, each with their own typical statistical behaviour, in which case the smoothness of PSR 1642−03 might be attributed to an averaging over many components which emit simultaneously; in contrast PSR 0950+18 would contain only a few components, so that there was a large chance that the emission would be zero from all components simultaneously. However, there is no indication as yet that any such statistical model is applicable.

9.5 Pulse drifting

The organised drift of sub-pulses seen in PSR 1919+21 (Fig. 9.4) was first noted by Drake & Craft (1968). Both in this pulsar, and in PSR 2015+28 which was studied by the same authors, the sub-pulses of successive pulses tend to occur at earlier phases, so that they drift fairly uniformly across the profile. They are followed by other sub-pulses which recur at fixed intervals of phase, two or more sub-pulses sometimes being present in one pulse. An idealised scheme is shown in Fig. 9.9.

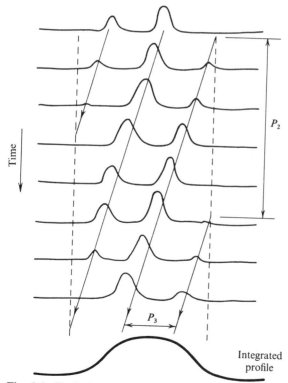

Fig. 9.9. Typical drifting sub-pulses. Successive pulses appear at the fundamental period P_1. The pattern repeats at interval P_2. Sub-pulses are separated by a typical interval P_3.

Individual radio pulses

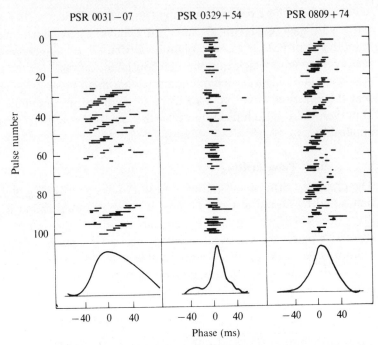

Fig. 9.10. Drifting and nulling. Each horizontal line is centred on the expected arrival time, with time increasing downwards and to the right. The positions of each sub-pulse are shown. PSR 0031−07 and PSR 0809+74 are typical negative drifters. PSR 0031−07 shows large null periods, missing about twenty pulses. (After Taylor & Huguenin, 1971.)

The drifting phenomenon is particularly well marked in PSR 0031−07. Here the average spacing between sub-pulses (P_3) is 55 ms and the drift rate is usually 8.5 ± 1.5 ms per period, or $3°.25$ per rotation. In this pulsar two other discrete drift rates are also observed for occasional periods of about 1 minute, giving drift rates of $1°.7$ per rotation and $5°.4$ per rotation respectively. At the usual drift rate the pattern recurs after about seven periods, so that the interval P_2 is about $6\frac{1}{2}$ s.

The most regular drifting behaviour so far observed is in PSR 0809+74. This pulsar emits more continuously than does PSR 0031−07, and the pattern of drifting can therefore be followed continuously. Cole (1970) showed that the drift rate varied, especially immediately after one of the breaks of emission which tend to occur at intervals of about 5 minutes (see Section 9.7 below). The breaks last for about six pulse periods. One such break can be seen in Fig. 9.10, where it is followed by a change both in phase and drift rate of the sub-pulse pattern. There is good

9.6 Modulation by pulse drifting

evidence that the pattern can be followed continuously through such breaks, which must then be regarded as incidental interruptions in a very stable drifting process. As a periodic phenomenon, the drifting process in this pulsar has a Q factor of over 1000.

Ritchings (1975) has found that the direction of drift is clearly differentiated between pulsars with high and low values of $\dot{P}$, as shown in Fig. 9.11.

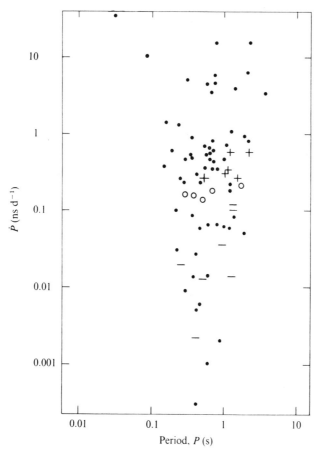

Fig. 9.11. Positive and negative drifters. All known drifters are marked $+$, $-$, or $\bigcirc$. Those marked $\bigcirc$ drift erratically. Other pulsars not known to show drifting are shown as dots on the diagram. (After Ritchings, 1975.)

9.6 Modulation by pulse drifting

The energy in single pulses depends on the phase of the sub-pulses as they drift across the profile. The pattern of sub-pulses repeats at intervals P_2 (see Fig. 9.9), so that the pulse energy is modulated at this periodicity. All

Individual radio pulses

drifting pulsars show this modulation; other pulsars also show some modulation at the same order of period, possibly due to some less obvious form of drifting behaviour.

A study of this modulation behaviour in twenty pulsars, and its relation to drifting, has been made by Taylor & Huguenin (1971). These authors obtained fluctuation spectra from autocorrelation functions of the pulse energies. Examples in Fig. 9.12 show:

(i) A very narrow spectral feature at 0.09 cycles per period in PSR 0809+74, as expected from the precision of the period P_2, which is close to $11P_1$.

(ii) A wider spectral line at 0.15 cycles per period for PSR 0031−07, the greater width corresponding to the more variable drift rate for this pulsar.

(iii) A narrow feature just below 0.5 cycles per period for PSR 0834+06. This feature, also seen in PSR 0943+10, represents a remarkable tendency for these pulsars to have alternate pulses strong and weak. More probably, this is a near coincidence in which the drift rate brings sub-pulses into the centre of the pulse in alternate pulses: this occurs if $P_2 \approx 2P_1$.

(iv) In many pulsars there is a rise in fluctuation spectrum at low frequencies which corresponds to a tendency for pulses to be emitted in groups or bursts. This is probably a distinct phenomenon from drifting. Both types of behaviour occur in some pulsars, e.g. PSR 1919+21 which shows low-frequency noise as well as a line near 0.25 cycles per period. This pulsar tends to show a modulation at $P_2 = 4P_1$, and also to emit pulses in groups of the order of twenty or thirty.

9.7 Nulling and moding

The histogram of pulse energy for PSR 0834+06 displays the tendency of this pulsar to miss out occasional pulses. The same effect of pulse 'nulling' is seen in several pulsars; for example Fig. 9.3 shows a missing pulse in PSR 1133+16. We have already noted the tendency for pulses to come in groups, which accounts for the enhancement of the low-frequency end of some fluctuation spectra (Fig. 9.12). These groups are often marked off by nulls, sometimes lasting several periods, which tend to occur at intervals characteristic of individual pulsars.

An example of pulse nulling in PSR 0809+74 is seen in Fig. 9.10. Here the interval between nulls is typically 200 rotation periods, i.e. about 250 s; the null lasts typically for five or six rotation periods. In PSR 0031−07, whose behaviour is in many ways similar to that of PSR 0809+74, the nulls occur at typical intervals of 100 rotation periods, i.e.

9.7 Nulling and moding

about 2 minutes. The nulls last as long as the active periods. The switching, on or off, occurs within the time of a single rotation of the pulsar.

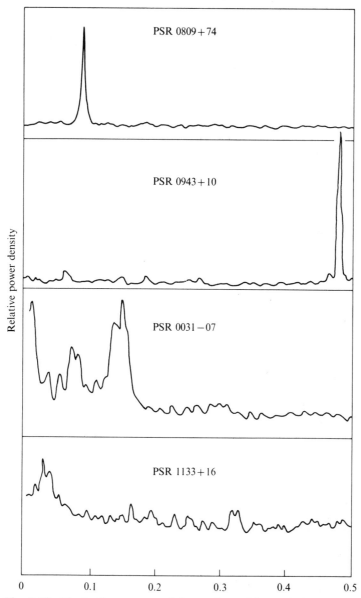

Fig. 9.12. Fluctuation spectra of four pulsars. The period P_2 is well defined in the 'drifting' pulsars PSR 0809+74 and PSR 0943+10. Drifting is more complex in PSR 0031−07, and irregular in PSR 1133+16. (After Taylor & Huguenin, 1971.)

Individual radio pulses

There is a clear distinction between the random variations of energy from pulse to pulse during the 'on' state, and the switch from 'on' to 'off' states. Pulse nulling can therefore be compared with the phenomenon of moding, which was described in the previous chapter as a variation in the integrated profile. The difference is that moding represents a switch between two configurations of the emitting region, apparently representing two different configurations of coherent particle motions, while nulling represents a complete stop to the radiation. The causes of the two phenomena may well be related, but there is as yet no understanding of the changes in the magnetosphere which act as the trigger between the two stable states in either phenomenon. Furthermore, the time scales involved are both very hard to understand, since oscillations and relaxation processes in a neutron star or in its magnetosphere generally have a time scale of less than a millisecond rather than some hundreds of seconds. The nulling phenomenon is commoner in the long-period pulsars, and it may be that it is a sign of old age. In Chapter 20, which discusses the question of pulsar ages, it is suggested that the cessation of radio pulses depends on the parameter $P\dot{P}^{-1/5}$, which marks a boundary in the plot of period P against its rate of change $\dot{P}$. The pulsars which show the most marked nulling lie just inside this boundary; this confirms both the existence of a definite boundary and the idea that nulling is an approach to a catastrophic final end to pulse emission.

References

Clark, R. R. & Smith, F. G. (1969). *Nature, Lond.* **221**, 724.
Cole, T. W. (1970). *Nature, Lond.* **227**, 788.
Cordes, J. M. (1975). *Astrophys. J.* **195**, 193.
Drake, F. D. & Craft, H. D. (1968). *Nature, Lond.* **220**, 231.
Hankins, T. H. (1971). *Astrophys. J.* **169**, 487.
Hankins, T. H. (1972). *Astrophys. J.* **177**, L11.
Hankins, T. H. & Rickett, B. J. (1975). *Methods in Comput. Phys.* **14**, 55.
Huguenin, G. R., Taylor, J. H. & Troland, T. H. (1970). *Astrophys. J.* **162**, 727.
Lyne, A. G., Smith, F. G. & Graham, D. A. (1971). *Mon. Not. R. astron. Soc.* **153**, 337.
Manchester, R. N., Tademaru, E., Taylor, J. H. & Huguenin, G. R. (1973). *Astrophys. J.* **185**, 951.
Ritchings, R. T. (1975). *Mon. Not. R. astron. Soc.*, in press.
Schonhardt, P. & Sieber, W. (1973). *Astrophys. Lett.* **14**, 61.
Smith, F. G. (1970*a*). *Mon. Not. R. astron. Soc.* **149**, 1.
Smith, F. G. (1970*b*). *Nature, Lond.* **228**, 913.
Smith, F. G. (1973). *Mon. Not. R. astron. Soc.* **161**, 9P.
Taylor, J. H. & Huguenin, G. R. (1971). *Astrophys. J.* **167**, 273.
Taylor, J. H., Manchester, R. N. & Huguenin, G. R. (1975). *Astrophys. J.* **195**, 513.

10
The Crab Nebula

10.1 Discovery and early observations

The Crab Nebula occupies a central place in the story of pulsars. Not only is it a remarkable and unique object in itself, presenting to us the most detailed picture of a young supernova remnant; it also contains the youngest and most energetic pulsar, which is the only pulsar known to radiate from long radio wavelengths through the whole electromagnetic spectrum up to hard X-rays. The nebula and the pulsar are both endlessly interesting in themselves; but their interaction opens up a new field in astrophysics, since it demonstrates a new form of energy transfer between a condensed body and a diffuse gas.

As a nebula, the Crab was first observed in 1731 by John Bevis, an English physicist and amateur astronomer. It took first place in the catalogue of nebulae compiled by Charles Messier in 1758, where it appeared as the nebula M1. The name 'Crab Nebula' was given to it about a hundred years later, when better telescopes revealed its tentacle-like structure. The present-day interest in the Crab Nebula dates mainly from the work by Baade in 1942, when he presented observations of its detailed structure and suggested that a prominent star near the centre of the nebula might be related to its origin. Baade already knew that the nebula was very young on an astronomical time scale. In 1939 Duncan had shown that the nebula was expanding at such a rate that it appeared to have originated in a point source only about 766 years earlier. But the most spectacular evidence of its youth was obtained from ancient Chinese and Japanese astronomical records, which described the appearance of a bright new star in the right part of the sky in the year AD 1054.

The extensive material in ancient and mediaeval Chinese records of comets and novae is described by Ho Peng-Yoke (1962). Records of a nova in the year 1054 were noted by Lundmark (1921), but the association with the Crab Nebula seems to be due to Duyvendak (1942). The history of the Sung Dynasty (Sung Shih, completed in 1345) contains this record: 'On a chi-chou day in the fifth month of the first year of the Chih-Ho reign-period a "guest star" appeared at the SE of Thien-Kuan (Taurus), measuring several inches. After more than a year it faded

The Crab Nebula

away.' The date is well corroborated in other independent records. The new star was visible in the day-time for several days, and remained as an object visible with the naked-eye at night-time for nearly 2 years. There is no doubt that the ancient records describe a supernova explosion, and the near-coincidence of positions and dates makes the identification with the present-day Crab Nebula entirely certain. The discrepancy between the actual birth date of AD 1054 and the date obtained by projecting back the presently measured velocities, which now converge at AD 1140 ± 10, is to be interpreted as a small but definite acceleration of the outward velocities, a fact of great significance in the question of the energy supply to the nebula.

Near the centre of expansion there are two stars, of 15th and 16th magnitude, which show prominently on good photographs of the nebula. Since there must be somewhere within the nebula a supply of energy to account for the continued emission of light, it was supposed that one of these two stars was the source of excitation. The one nearest to the centre of expansion was the south preceding star; it was also a star with a most unusual spectrum. Baade suggested that this might be the parent star for the whole nebula, but he was unable to account for the excitation through the familiar process of ultraviolet light emission. The spectrum showed no emission or absorption lines, which suggested that the star had a very high temperature. But at the same time there was no indication of the abnormally low colour index which would correspond to the strong ultraviolet emission of a hot star. We know now that this star is the Crab Pulsar, and that it feeds energy into the nebula not by light but through high-energy particles accelerated in a rotating magnetic field.

The Crab Nebula is contained within an ellipse 180×120 arc seconds across. The outer parts are filamentary, forming a network enclosing the more luminous central part. This is an amorphous mass concentrated towards the centre but extending over most of the major diameter and about two-thirds of the minor diameter. The light from these two components is totally different in character, so that in a colour photograph of the nebula the filaments show as predominantly red, while the centre is white or bluish-white. The red light from the filaments is line radiation, mainly $H\alpha$ but including many other lines such as [NII], [OI], [OII], [OIII], [SII], [NeIII], HeI and HeII. (The square brackets indicate forbidden transitions.) The relative abundance of the elements is close to the standard solar composition, except for a rather higher helium abundance. The ionisation is due to the ultraviolet light from the nebula.

The line radiation from the centre of the nebula originates in filaments on the front and back, i.e. the parts that approach and recede from the

10.2 Continuum radiation from the Crab Nebula

observer with maximum velocities. The Doppler shifts in these lines correspond to expansion velocities close to 1000 km s^{-1}. By combining this value for the expansion in the line of sight with the measured angular rate of expansion, and assuming that the expansion follows a simple elliptical form, the distance of the nebula may be obtained. This measurement and other estimates of the distance place the nebula in the range 1.5 to 2.5 kpc.

The white light from the central amorphous component has no spectral lines, and its origin was for a time a complete mystery. Although this component usually appears to be amorphous, under good seeing conditions it is found to be concentrated in fine filaments, like cotton wool. These fibrous concentrations run in organised directions, which are now known to be associated with a magnetic field within the nebula. The spectrum, and the high brightness, of this source of continuous radiation are incompatible with thermal radiation.

10.2 The continuum radiation from the Crab Nebula

In 1949 the radio astronomers J. G. Bolton, G. J. Stanley and O. B. Slee identified the Crab Nebula as a radio source. This was the first identification of a galactic radio source. The radio flux density was greater than in visible light, so that it became even more difficult to explain the continuum radiation in the familiar terms of thermal radiation from ionised gas.

The explanation of this bright continuum radiation was provided by I. Shklovsky in 1953. High-energy electrons moving in a magnetic field follow curved paths; this curvature implies an acceleration, which leads to radiation. Previous analyses of energy loss in a synchrotron electron beam had already shown that this was an important effect, and it is generally called synchrotron radiation. It is also known as magnetic braking radiation, magnetobremsstrahlung. The main characteristics of synchrotron radiation are outlined in Chapter 15. The importance of this suggestion was that it provided the only known means for a very hot gas to radiate efficiently and over a wide range of wavelengths. Furthermore it led to the prediction that the radiation at any wavelength would be at least partly linearly polarised.

Confirmation of the synchrotron proposal soon came from observations by two Soviet astronomers, Vashakidze (1954) and Dombrovsky (1954), who showed that there was indeed a large linear polarisation. A detailed investigation by Oort & Walraven (1956) is reported in a classic paper. Photographs in this paper show that the polarisation is so high that the detailed appearance of the nebula varies dramatically according to

The Crab Nebula

the setting of a Polaroid filter on the telescope. The integrated light from the whole nebula is 9% polarised, while locally the polarisation may reach 60%. Oort & Walraven showed that the white radiation from the nebula must indeed be synchrotron radiation; their analysis showed that the magnetic field strength in the nebula must be about 10^{-3} gauss, and the electron energies must extend at least up to 10^{11} eV. The radiation mechanism was now understood, but the origin of the magnetic field and of the very high electron energy was to remain a mystery until the discovery of the Crab Pulsar.

Radio observations of the Crab Nebula now extend over the wavelength range from 30 m to 3 mm. The optical observations have been extended into the infrared, covering 500 nm to 5000 nm. X-ray and gamma-ray observations are now made from rockets and satellites, over the energy range 0.5 keV to 500 keV, corresponding to a wavelength range 3 nm to 3 pm. The known spectrum therefore covers a range of $10^{13}:1$, or 43 octaves, spanning frequencies from 10^7 to 10^{20} Hz. A review by Baldwin (1971) shows that the spectrum is probably continuous, with the spectral indices and flux densities in the three main parts of the spectrum as shown in Table 10.1.

TABLE 10.1 *Crab Nebula spectrum*

	Range	Index $\alpha(S=\nu^\alpha)$	Flux density (W m^{-2} Hz^{-1})	Frequency ν (Hz)
Radio	10^7–10^{14}	-0.26	10^{-23}	10^{12}
Optical	5×10^{12}–5×10^{13}	-1.0	10^{-25}	10^{14}
X-ray	10^{17}–10^{20}	-1.2	10^{-29}	10^{18}

10.3 The energy supply

The analysis by Oort & Walraven of the optical and radio emission led to a fairly precise definition of the energy spectrum and actual numbers of electrons within the nebula, as well as the average value of the magnetic field. The total energy of fast particles in the nebula was found to be of the order of 10^{49} erg, most of this being concentrated in particles with energy of order 10^{11} eV. This energy is about one-thousandth of the total energy that would be released if a solar mass burned from hydrogen into helium, which is the maximum amount of nuclear energy that could reasonably be expected from a supernova explosion. It would be remarkable, but not inconceivable, for the energy of the explosion to be so well concentrated

10.3 The energy supply

into high-energy particles. But a further problem was pointed out by Oort & Walraven. The electrons must be radiating so efficiently that their lifetimes are only of order 100 years rather than 1000 years, so that they could not have been accelerated in the original explosion.

The lifetime of an electron radiating synchrotron radiation with a maximum spectral density at frequency ν(Hz) in a field B(gauss) is expressed as a half-life:

$$t_{1/2} = 10^{12} \nu^{-1/2} B^{-3/2} \text{ seconds}. \tag{10.1}$$

The lifetime could only be extended for optical radiation at a fixed frequency by assuming a smaller value of the field B, which would imply a larger total electron energy to produce the observed emission. The total energy would then reach or exceed the total available from the supernova explosion. This dilemma was made far worse by the observation of X-ray emission at energies in excess of 100 keV, where the synchrotron radiation must have come from electrons with energies of at least 10^{14} eV, which would have lifetimes of less than a year in any reasonable magnetic field.

An equally difficult problem is presented by the existence of the magnetic field itself, which contains the same order of magnitude of energy as the particles. Although there is no loss through radiation, it is impossible that this field is merely a remnant of a field which simply originated at the time of the supernova explosion. Any such field would have been reduced far below 10^{-3} gauss in an adiabatic expansion, transferring most of its energy into expansion energy of the nebula. There must be a means for the continued generation of a magnetic field throughout the nebula.

There was, therefore, even before the discovery of the Crab Pulsar, incontrovertible evidence that an energy source existed within the Crab Nebula that was providing both the high-energy particles and the magnetic field throughout the nebula. The location of this source was suspected to be at, or close to, Baade's star, both on account of its unusual spectrum and because of some remarkable activity in the nebula close to the star.

The activity close to Baade's star had been noted in 1921 by Lampland. Later observations, and especially some by Scargle & Harlan (1970), confirmed his suggestion that some nebulous wisps about 10 arc seconds away from the star were moving and changing in brightness, sometimes even within periods of only a few months. If these were bulk movements of material, they must be moving at speeds approaching the speed of light. Between the wisps and the star there seemed to be a relatively empty space.

The Crab Nebula

10.4 The transfer of energy from the pulsar to the nebula

In 1957 Piddington suggested that the magnetic field of the nebula might originate in a massive rotating body within the nebula. The field round a rotating magnetised star may be distorted by a conducting atmosphere; if there is an outward flow of ionised plasma, in the form of a stellar wind, the magnetic field lines may be trapped and extended outwards radially. Another component of the field may be predominantly toroidal, corresponding to a radial outflow of charge. Piddington showed that the field round a rotating source will be forced into a spiral, like the spiral arms of a galaxy, and that the field strength will be enhanced in the process. The 'winding-up' process will give a field strength increasing linearly with time, unless the nebula is at the same time expanding, or unless the field configuration is tending to break down, releasing its energy by accelerating charged particles.

The idea that the particle energy in the Crab Nebula also came from a rotating magnetic star was due to Pacini (1968), who showed that a neutron star could provide sufficient energy from its rotational energy to account for the whole present-day luminosity of the nebula. As we have seen in Chapter 3, this was a remarkably perceptive view, published a year in advance of the discovery of the Crab Pulsar. With the source established, it is now possible to look back at the evolution of the nebula (Pacini & Salvati, 1973) and ask whether the whole of the field and particle energies have been generated since the supernova explosion, or whether some at least of the low-energy electrons might have existed for the whole life of the nebula.

The magnetic energy W_B is derived from the rotational energy of the star, which gives a supply L depending on the time t as:

$$L = L_0 \left(1 + \frac{t}{\tau}\right)^{-\alpha} \tag{10.2}$$

where τ is the time scale of the rotational slowdown. The index α is related to the braking index (Chapter 7) by:

$$\alpha = \frac{n-2}{n-1}. \tag{10.3}$$

For purely magnetic dipole radiation $n = 3$ and $\alpha = 2.0$; although the observational evidence does not correspond exactly with $n = 3.0$, the general rule is near-enough correct. The field also loses energy through adiabatic expansion; if the expansion velocity is v, and the radius R, there

10.4 Energy transfer from pulsar to nebula

is a total net gain according to

$$\frac{d}{dt}W_B = L_0\left(1+\frac{t}{\tau}\right)^\alpha - W_B\frac{v}{R}. \tag{10.4}$$

Pacini & Salvati traced the history of the field energy by integrating this equation, and showed that the original field around the supernova is rapidly submerged in the 'wound-up' field, which reaches a peak of several hundred gauss about 10 days after the explosion. Thenceforward the field declines, with a strength of order $10^8 \, t^{-1}$ gauss at time t (in seconds). In this strong initial field all particles would lose so much energy that they could not radiate even at the lowest radio frequencies. The whole of the present field, and the high-energy electrons, must therefore be sustained by a continuing process.

Two stages may be distinguished in the transfer of energy from the pulsar to the nebula. Close to the pulsar, and out to a boundary which extends to a radius R_s, which is about one-tenth of the radius of the nebula, there is an outward stream of field and particles which is unaffected by the presence of the nebula. The energy of the stream may be predominantly contained in the particles, as in the Goldreich & Julian model of an aligned rotator (Chapter 6), or in the radiating magnetic field of the orthogonal rotating dipole discussed by Pacini; probably both are important. Rees & Gunn (1974) have discussed the second stage, when the 30-Hz radiation field of the rotating dipole reaches the boundary of the cavity. Tentatively one may identify this boundary with the active wisps of radiation discovered by Lampland. The 30-Hz radiation field is absorbed in the boundary, transferring energy to relativistic electrons.

The magnetic field in the nebula is then supposed to be generated as a toroidal field related to the stream of particles out from the pulsar. This is not an oscillatory field, and it is not absorbed at the boundary. Instead it is spread through the rest of the nebula, with a field strength depending inversely on the expansion velocity. The velocity falls as radius increases, so the field tends to increase towards the edge of the nebula.

This model accounts well for the general properties of the amorphous part of the nebula. The position of the cavity boundary can be found by assuming that the outward radiation and particle pressure $L/4\pi c R_s^2$ are balanced by the energy density of the magnetic field and particles outside the cavity. Then if the energy outside the cavity has all accumulated during a time of the order of the age of the nebula, and the nebula is expanding at a rate $\dot{R}_N$, the ratio of R_s to the nebula radius R_N is found by

The Crab Nebula

Rees & Gunn to be:

$$\frac{R_s}{R_N} = \left(\frac{\dot{R}_N}{c}\right)^{1/2}. \tag{10.5}$$

This gives a reasonable value for R_s equal to about one-tenth of R_N.

The detailed processes at the boundary are probably very complex. However, it is clear that 30-Hz radiation will be absorbed by an inverse of the synchrotron process, since the nebula is optically very thick ($\tau \approx 10^6$) for 30-Hz synchrotron radiation. It is also certain that high-energy particles will transfer their energy to the nebula by plasma processes. Kahn (1971) has shown how Alfven waves with a range of wavelength can be excited by a stream of relativistic electrons. These waves serve to redistribute the energy among a larger number of particles.

Although much of the discussion is necessarily in general terms, there is no doubt that:

(i) the Crab Pulsar is the source of energy for the continuing luminosity of the nebula;
(ii) energy can be fed from the pulsar into the nebula both in the form of 30-Hz waves and as a particle flux;
(iii) the acceleration of the expansion occurs as a result of the continued injection of energy from the pulsar.

As will be seen from the next chapter, the detailed behaviour of the pulsar itself is not as easy to understand.

References

Baldwin, J. E. (1971). *IAU Symposium No. 46*, p. 22. (Dordrecht: D. Reidel.)
Dombrovsky, V. A. (1954). *Dokl. Akad. Nauk USSR* **94**, 1021.
Duyvendak, J. J. L. (1942). *Publ. astron. Soc. Pacific* **54**, 91.
Ho Peng-Yoke (1962). *Vistas in Astronomy*, ed. A. Beer, p. 127. (Oxford: Pergamon Press.)
Kahn, F. D. (1971). *IAU Symposium No. 46*, p. 281. (Dordrecht: D. Reidel.)
Lampland, C. O. (1921). *Publ. astron. Soc. Pacific* **33**, 79.
Lundmark, K. (1921). *Publ. astron. Soc. Pacific* **33**, 225.
Oort, J. H. & Walraven, Th. (1956). *Bull. astron. Inst. Neth.* **12**, 285.
Pacini, F. (1968). *Nature, Lond.* **219**, 145.
Pacini, F. & Salvati, M. (1973). *Astrophys. Lett.* **13**, 103.
Piddington, J. H. (1957). *Aust. J. Phys.* **10**, 530.
Rees, M. J. & Gunn, J. E. (1974). *Mon. Not. R. astron. Soc.* **167**, 1.
Scargle, J. & Harlan, E. (1970). *Astrophys. J.* **159**, L143.
Shklovsky, I. S. (1953). *Dokl. Akad. Nauk USSR* **90**, 983.
Vashakidze, M. A. (1954). *Astr. Circ. No. 147*.

11
The Crab Pulsar

11.1 The spectrum

The discovery of the Crab Pulsar, first as a source of radio pulses and later as a source of light pulses, came many years after the first observations of the unusual object at the centre of the nebula. Optically, the first observations date back to the earliest photographs in which the central pair of stars was resolved, although the problem of interpreting the spectrum of the south-preceding star (known until recently as Baade's star) was not discussed until the papers by Baade and Minkowski were published in 1942. There were also some early radio observations at long radio wavelengths, which still represent one extreme end of an extraordinarily wide spectrum over which the pulsar can be detected. In these radio observations it became clear only shortly before the first pulsar was discovered that the existence of such a compact and intense source of emission was hard to explain in the usual astrophysical terms of synchrotron emission (Bell & Hewish, 1967). Neither in the radio nor in the optical domain was there any evidence that this intense source might be pulsating so rapidly that the limited time resolution of the detectors was giving a false impression of a steady signal.

In fact, at the time of the discovery of radio pulses from the Crab Pulsar, a recording of X-ray pulses already existed, from equipment with good time resolution flown on a balloon in 1967 (Fishman, Harnden & Haymes, 1970). Again, at the time of the discovery of optical pulses, a similar recording had already been made, but not analysed, by Willstrop (1971) in a search for rapid periodic fluctuations of light both in known radio pulsars and in supernova remnants. It was hard to realise at first that this astonishing object was giving precisely periodic pulses over the whole available electromagnetic spectrum.

Detailed observations of the pulsar now extend throughout the spectrum from 20-MHz radio waves to 600-kV gamma-rays, almost the same range over which the nebula itself is known to radiate. The radiation from the nebula forms a continuous spectrum, which is easily accounted for as synchrotron radiation. The pulsar spectrum is in two distinct parts; the radio spectrum falls steeply to undetectable levels at a few gigahertz,

The Crab Pulsar

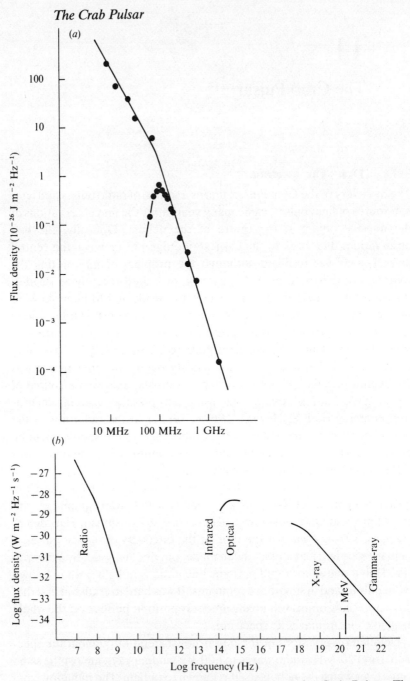

Fig. 11.1. (a) Radio frequency spectrum of the Crab Pulsar. The apparent pulse energy falls below 100 MHz due to scintillation, but the power spectrum continues to rise towards lower frequencies. (b) Spectrum of the Crab Pulsar covering 40 octaves, from radio to gamma-rays.

11.1 The spectrum

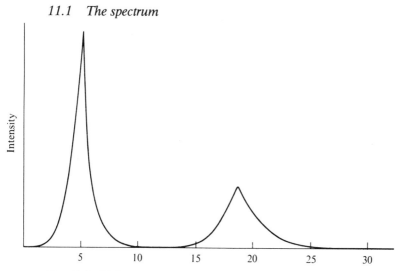

Fig. 11.2. Light curve of the Crab Pulsar. (200-inch telescope; after Visvanathan, 1971.)

while the optical, X-ray and gamma-ray spectra are continuous and apparently quite separate from the radio spectrum. Fig. 11.1 shows the spectrum of the mean flux density.

Three areas of uncertainty in this spectrum require comment. At radio frequencies below 50 MHz the pulses are smeared out by interstellar scattering, so that it is an assumption that the source is still the same pulsating object. This actually only involves the last 2 octaves in the total of 40 octaves, but there is a good indication of its correctness in the very high brightness at these long wavelengths, which could not be accounted for in any other known type of source. Second, at infrared wavelengths the spectrum appears to be turning down in comparison with the visible range. This is evidently important, since it may be interpreted in terms of self-absorption, but it must be remarked that calibrations of infrared observations are difficult and the precise form of this part of the spectrum may still be in doubt.

Finally, the gamma-ray spectrum is fairly well established up to 600 kV. There are some observations above this energy which need further confirmation. Spark chamber observations, using balloons, have given positive measurements at minimum energies of 10 MeV. The most sensitive measurements (Schonfelder, Lichti & Moyano, 1975) provide only upper limits for gamma-rays in the range 1–10 MeV. These upper limits lie approximately on an extrapolation of the spectrum in Fig. 11.1. Attempts to extend the spectrum to cosmic-ray energies around 10^{12} eV by observing Cerenkov light from air showers, and searching for a 33-ms

The Crab Pulsar

periodicity, have led to a positive result which remains unconfirmed (Grindlay, 1971).

11.2 Pulse shapes

The integrated pulse profile of the Crab Pulsar is very similar over the whole spectrum, except at long radio wavelengths where it is distorted by interstellar scattering. Fig. 11.3 shows the profiles for radio, optical, and X-rays. The main features to note are:

(i) There is little or no difference over the whole of the visible spectrum, i.e. the colour of the light is the same throughout the profile.

(ii) In the X-ray and gamma-ray region the ratio of the main pulse to interpulse energies changes slowly, so that for a photon-energy of about 1 keV the energies are equal, instead of in the ratio of 2:1 as in the visible spectrum.

(iii) Between the main pulse and interpulse a continuous bridge of emission appears in X-rays, and less strongly in the visible spectrum.

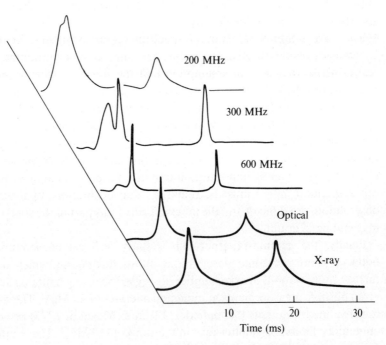

Fig. 11.3. Integrated pulse profiles for the Crab Pulsar, from X-rays to radio frequencies.

11.3 Dispersion measure and pulse arrival times

(iv) At short radio wavelengths the profiles of both the main pulse and the interpulse are narrower than in the visible spectrum; note, however, that the sharp cusp at the peak of the visible curve matches the sharpness of the radio peak.

(v) At 18-cm wavelength, the shortest radio wavelengths so far used, the interpulse is much less intense. Manchester (1971) shows that the spectral index between 1664 MHz and 410 MHz is -3.7 for the interpulse and -2.8 for the main pulse.

(vi) At radio wavelengths greater than 30 cm (frequency 1 GHz) a third component appears, immediately preceding the main pulse. This is known as the precursor (see Fig. 11.3). It has a steeper spectrum than the rest of the radio emission, falling with frequency above 400 MHz at least as ν^{-5} as compared with $\nu^{-3.3}$ for the whole pulsar energy. At 610 MHz the precursor pulse energy is approximately 0.1 of the main pulse energy.

(vii) The pulse smearing seen at wavelengths greater than 1 m is due to multipath scattering in the interstellar medium (Chapter 14). The effect is variable: in 1974 it increased so much that most of the pulse structure was obscured even at 408 MHz (Lyne & Thorne, 1975).

11.3 Dispersion measure and pulse arrival times

The dispersion in pulse arrival times may be measured with great accuracy at radio wavelengths, due to the sharpness of the integrated pulse profile. The value may be used to predict the arrival time for visible light and X-rays. Even though entirely different techniques and locations are used for detecting the pulses in the three regimes, it has been possible to show that the pulses must have been transmitted within a time interval of 300 μs. The sources of the radio, light and X-rays must therefore be within 100 km of each other.

The dispersion measure itself is found to vary (Fig. 11.4). The peak variation is 0.1% of the total, corresponding to a change in electron content along the line of sight amounting to 2×10^{18} electrons cm^{-3}. It seems likely that this variation is within the nebula, although there is no direct information on the thermal electron density (but see Chapter 13 on Faraday rotation). There is a strong suggestion that one of the large changes in dispersion measure occurred at the time of a glitch, in June 1969, so that the discontinuity might be associated with conditions very close to the pulsar. The problem in this association is that all electrons close to the pulsar, which might be affected by the glitch, are expected to have high relativistic energies, so that they will not contribute fully to refractive index effects.

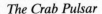

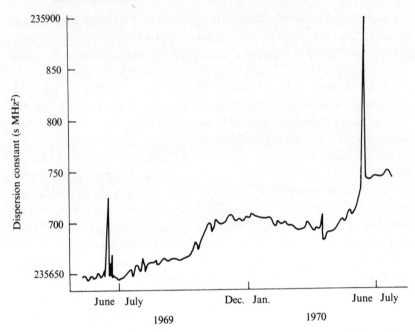

Fig. 11.4. Variations in dispersion measure for the Crab Pulsar. (After Rankin & Roberts, 1971.)

11.4 Variations of intensity

As with all other pulsars, the radio emission from the Crab Pulsar is variable on a wide range of time scales. A long-term variation over periods of months and years has been followed by Rankin and by Lyne & Thorne (1975); the range of variation covers a factor of more than two, and all components are observed to vary together.

The optical intensity, in contrast, is essentially constant. The brightness of Baade's star, as compared with its companion star on a series of photographs, has not varied by more than 0.2 mag over some tens of years. Photometric measurements over the last few years show less than 0.05 mag variation (Kristian, 1971).

11.5 Variability of pulses: giant pulses

There are no variations from pulse to pulse in the light from the Crab Pulsar, as there are in the radio pulses from this and all other pulsars. The accuracy for light pulses is, however, limited necessarily by photon noise, since a single optical pulse received in a large telescope may typically contain only ten photons.

11.5 Variability of pulses: giant pulses

A test of the variability of optical pulses was made by Hegyi, Novick & Thaddeus (1971). The intensity of each pulse was recorded as a photon count N, and the two averaged quantities $\langle N \rangle^2$ and $\langle N^2 \rangle$ were compared. Over a typical averaging time of 5 min no difference was observed in the ratio of these two quantities at any part of the main pulse profile. This excluded the possibility of fluctuations similar to the pulse-to-pulse variations familiar in radio pulses from other pulsars.

In contrast to the constancy of the light pulses, the intensity of the radio pulses from the Crab Pulsar is more variable on a short time scale than for any other pulsar so far investigated (see Chapter 9). Unfortunately most of the individual pulses are submerged in the background noise of the nebula, which can only partly be smoothed out during the very short duration of the pulse. The histogram of pulse intensity in Fig. 11.5 therefore shows a large random component centred on zero, due to noise, while only the larger pulses contribute effectively to the histogram. The remarkable part of the histogram lies in the long tail extending to large pulse intensities. These are so large that the flux density in the pulse, assuming the pulse duration to be that of the integrated profile, would exceed the flux density of the entire Crab Nebula.

The 'giant pulses' occur on average about once every 5 or 10 minutes, i.e. one in every 10^4 pulses is a giant. The interval between them is

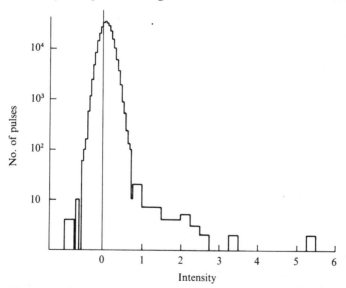

Fig. 11.5. Histogram of pulse intensity for the Crab Pulsar observed at 160 MHz. The peak around zero intensity is mainly due to receiver noise, and the pulses are seen as an exponential tail of higher intensities. (After Sutton, Staelin & Price, 1971.)

The Crab Pulsar

randomly distributed. Giant pulses occur at the time either of the main pulse and the interpulse, but apparently not at the time of the precursor. Some observers consider that they represent a distinct kind of pulse, although there seems no reason to regard the histogram of Fig. 11.5 as different from a continuous distribution.

T. Hankins (Hankins & Rickett, 1975) has attempted to resolve the structure of giant pulses observed at 430 MHz, by using a large radio bandwidth with a de-dispersion technique (Chapter 2). Using a total bandwidth of 2 MHz and a resolution time of 40 μs, he found that the pulses were very much narrower than the integrated radio profile, and consequently very much more intense than had been apparent. A single pulse recorded on 30 June 1973 reached a flux density of 86 000 Jy, which is sixty times greater than the flux density of the Crab Nebula at this frequency. The rise time was unresolved; it was followed by an exponential decay with time scale 90 μs, which appears to be a genuine decay time for the source since it is longer than the expected time scale of interstellar pulse broadening, and it is well resolved by the observing technique.

The sharp time of onset of the giant pulses is distributed over ± 200 μs from the centres of the main pulse and the interpulse. In this respect they resemble the sub-pulses of other pulsars, which appear at times distributed through the broader integrated pulse profiles.

The giant pulses have a narrower spectrum than might be expected from the continuum spectrum of the total pulsar radiation. Observations by Heiles & Rankin (1971) showed little correlation between the strength of giant pulses observed at 74, 111 and 318 MHz. There is some reason to believe that the radiation in these pulses is essentially narrow-band, as in a resonance phenomenon (Chapter 18).

11.6 Polarisation: optical

The optical pulses from the Crab Pulsar show linear polarisation. This is of course only detected on integration over many pulses, but it is reasonable to suppose that the lack of intensity variation from pulse to pulse indicates that the variability of polarisation typical of radio pulses is not present in the optical pulses. We therefore refer only to 'the optical pulse', although in interpretation one must bear in mind that this may represent an integration over an extended source rather than a single polar diagram of emission (Chapter 17).

The linear polarisation changes in degree and position angle through the pulse. The observed linear polarisation includes a component due to interstellar scattering, which must be subtracted to obtain the original source polarisation. When the source polarisation is low, the results of

11.6 Polarisation: optical

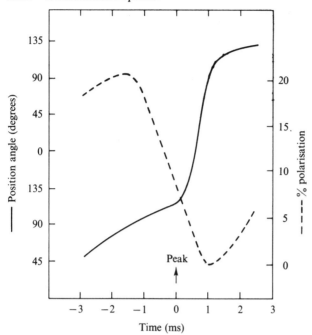

Fig. 11.6. Polarisation of light from the Crab Pulsar. One millisecond after the main pulse peak the percentage polarisation falls to zero, and the position angle swings through 180°. (After Cocke et al., 1970.)

this subtraction depend critically on the interstellar component. Fig. 11.6 presents the polarisation measurements of Cocke, Disney, Muncaster & Gehrels (1970) in the form of plots of percentage polarisation. The average interstellar polarisation for ten stars within a few minutes of arc of the Crab Nebula was found to be $2.0 \pm 0.2\%$ at position angle $147° \pm 3°$; this appears in a vector plot as a displaced origin from which the true polarisation can be found. The correction has little effect at the beginning and end both of the main pulse and the interpulse, but at the centres of each pulse where polarisation is low the changes in position angle may easily be misinterpreted if the interstellar polarisation is wrongly measured.

The resultant percentage polarisation and position angles are shown in Fig. 11.7. Similar results were obtained by Wampler, Scargle & Miller (1969) and by Kristian, Visvanathan, Westphal & Snellen (1970), but with less certainty in the interstellar correction.

The notable features of the optical polarisation are:

(i) The high degree of linear polarisation, especially at the start of the main pulse, where it exceeds 20%;

The Crab Pulsar

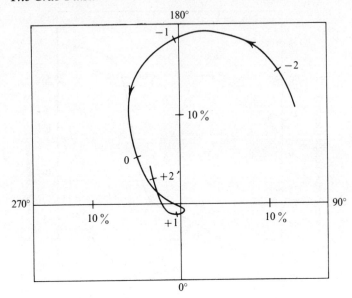

Fig. 11.7. Polarisation of light from the Crab Pulsar. The data of Fig. 11.6. Numbers on the curve represent times in milliseconds relative to the main pulse peak.

(ii) the smooth variation of percentage polarisation through the main pulse, falling to less than 1% soon after the peak intensity and then rising again;
(iii) the swing of position angle through at least 150°;
(iv) the similarity of behaviour in the main pulse and the interpulse.

A special search for circular polarisation with a time resolution of 2.65 ms gave only an upper limit of about $\frac{1}{2}$% at the centres of the main pulse and the interpulse (Cocke, Muncaster & Gehrels, 1971).

11.7 Polarisation: radio

As for other pulsars, a clear distinction must be made between integrated pulse profiles and individual radio pulses. There are, however, very few observations of the polarisation of individual pulses from the Crab Pulsar, and these are of doubtful validity since they can only refer to giant pulses which, as described in Section 11.5 above, have a very short time scale. Graham, Lyne & Smith (1970) showed that some giant pulses observed at 408 MHz were highly polarised, while Argyle (1973) showed there is almost no polarisation observable at 146 MHz where the pulses are appreciably broadened by interstellar scattering. No observations have

11.8 Relation between optical and radio radiation

yet been made which would resolve both the time structure and the polarisation of giant pulses.

The average polarisation, i.e. the result of integrating the Stokes parameters to produce an integrated pulse profile, was measured at 410 MHz by Campbell, Heiles & Rankin (1970). The precursor is very highly linearly polarised, possibly almost 100%, with a position angle which does not rotate appreciably. The main pulse and the interpulse are much less polarised, and it has been very difficult to measure the polarisation characteristics, especially at low frequencies where the pulses are smeared out by interstellar scattering. There is no indication of a dip in percentage polarisation near the centre of the radio pulse, as there is optically.

Apart from the precursor, which cannot be detected at high radio frequencies, there is no difference in the polarisation at frequencies up to 1664 MHz (Manchester, 1971). At low frequencies the components of the pulse profile become broadened by interstellar scattering and polarisation is not seen so clearly. At 150 MHz Schonhardt (1971) found that the precursor and the main pulse blended together to give 20% linear polarisation, while at 111 MHz Heiles, Rankin & Campbell (1970) found less than 20% linear polarisation. These results are not compatible with a blend in equal parts of an almost unpolarised main pulse and a fully polarised precursor, but it is unfortunately not possible to decide whether the precursor is merely weaker or whether it is less polarised.

11.8 Relation between the optical and the radio radiation from the pulsar

There are several characteristics which suggest that there is a basic distinction between the two sections of the spectrum shown in Fig. 11.1, i.e. between the radio region and the combined optical, X-ray and gamma-ray region.

(i) Extrapolation of the steep radio spectrum and the downturn in the infrared suggests a deep minimum in the millimetre wavelength range.

(ii) The strength of individual pulses varies in the radio range but not at optical wavelengths.

(iii) The polarisation behaviour is similar throughout the radio range but different at optical wavelengths.

The differences are sufficiently striking that it is all the more remarkable that the pulse shapes are basically very similar at all wavelengths. The giant radio pulses show, however, that it is the integrated radio pulse profile which is close to the optical pulse profile, not the individual pulses.

The Crab Pulsar

The theory of emission developed in Chapters 16 to 18 suggests that the integrated profile represents, for both parts of the spectrum, a spatial distribution of sources. The sharpness of the cusp at the peak of intensity then represents the narrowest feature, which may be identified with the beamwidth of radiation from a single source. Following this interpretation, the two parts of the spectrum originate in the same or closely adjacent places, but by different radiation mechanisms.

References

Argyle, E. (1973). *Astrophys. J.* **183**, 973.
Baade, W. (1942). *Astrophys. J.* **96**, 188.
Bell, S. J. & Hewish, A. (1967). *Nature, Lond.* **213**, 1214.
Campbell, D. B., Heiles, C. & Rankin, J. M. (1970). *Nature, Lond.* **225**, 527.
Cocke, W. J., Disney, M. J., Muncaster, G. W. & Gehrels, T. (1970). *Nature, Lond.* **227**, 1327.
Cocke, W. J., Muncaster, G. W. & Gehrels, T. (1971). *Astrophys. J.* **169**, L119.
Fishman, G. J., Harnden, F. R. & Haymes, R. C. (1970). *Astrophys. J.* **156**, L107.
Graham, D. A., Lyne, A. G. & Smith, F. G. (1970). *Nature, Lond.* **225**, 526.
Grindlay, J. E. (1971). *Nature, Lond.* **234**, 153.
Hankins, T. H. & Rickett, B. J. (1975). *Methods in Comput. Phys.* **14**, 55.
Hegyi, D., Novick, R. & Thaddeus, P. (1971). *IAU Symposium No. 46*, p. 129. (Dordrecht: D. Reidel.)
Heiles, C. & Rankin, J. M. (1971). *Nature Phys. Sci.* **231**, 97.
Heiles, C., Rankin, J. M. & Campbell, D. B. (1970). *Nature, Lond.* **228**, 1074.
Kristian, J. (1971). *IAU Symposium No. 46*, p. 87. (Dordrecht: D. Reidel.)
Kristian, J., Visvanathan, N., Westphal, J. A. & Snellen, G. H. (1970). *Astrophys. J.* **162**, 475.
Lyne, A. G. & Thorne, D. J. (1975). *Mon. Not. R. astron. Soc.* **172**, 197.
Manchester, R. N. (1971). *Astrophys. J.* **163**, L61.
Minkowski, R. (1942). *Astrophys. J.* **96**, 199.
Rankin, J. M. & Roberts, J. A. (1971). *IAU Symposium No. 46*, p. 114. (Dordrecht: D. Reidel.)
Schonfelder, V., Lichti, G. & Moyano, C. (1976). *Nature, Lond.* **257**, 375.
Schonhardt, R. (1971). *IAU Symposium No. 46*, p. 110. (Dordrecht: D. Reidel.)
Sutton, J. M., Staelin, D. H. & Price, R. M. (1971). *IAU Symposium No. 46*, p. 97. (Dordrecht: D. Reidel.)
Visvanathan, N. (1971). *IAU Symposium No. 46*, p. 152. (Dordrecht: D. Reidel.)
Wampler, E. J., Scargle, J. D. & Miller, J. S. (1969). *Astrophys. J.* **157**, L1.
Willstrop, R. V. (1971). *IAU Symposium No. 46*, p. 152. (Dordrecht: D. Reidel.)

12
The interstellar medium as an indicator of pulsar distances

12.1 The interstellar electrons

The frequency dispersion in pulse arrival times provides a precise measure of the total amount of ionised gas along the line of sight to each pulsar. This would in turn provide a measure of the distance of each pulsar, provided that the electron density within the electron gas were everywhere uniform and constant. A calibration of the distance scale would be available from the Crab Pulsar, which is known to be at a distance of 2 kpc. The dispersion measure $DM = 58$ pc cm^{-3} then gives a mean density $\langle n_e \rangle = 0.03$ cm^{-3}. For many pulsars the only possible estimate of distance is obtained from the dispersion measure and this single value of electron density. There is, however, good evidence that the electron density is not uniform, but that it varies through the Galaxy both on a large scale, reflecting the known distribution either of neutral gas or of ionising agencies, and on a small scale corresponding to the visible HII regions.

The large-scale distribution of the electron gas in the Galaxy is best considered in two components. The main component corresponds roughly to the hydrogen gas whose neutral component is observable through the 21-cm line. This is distributed as a disc, extending in radius far beyond the Sun. We shall initially disregard the spiral structure within this disc, and describe the electron distribution through the simple parameters of mean density on the galactic plane $\langle n_{e0} \rangle$ and a thickness parameter z_0.

The second component is a thinner disc, with radius less than the solar distance from the galactic centre. This comprises the highly ionised HII regions surrounding the bright O and B type stars found in the central disc of the Galaxy. We will see that observations of pulsar dispersion give useful information on both these components of the electron distribution, but we first review the evidence available from other sources.

Diffuse ionised gas in interstellar space may be detected in several different ways. At radio wavelengths the gas may have an appreciable optical depth due to bremsstrahlung, i.e. free–free collisions. There may consequently be thermal emission, or there may be absorption of radio

The interstellar medium

emission from other radio sources beyond the gas. These sources may be extragalactic or galactic and they may be discrete sources or a diffuse background. Radio astronomy also allows observation of the ionised gas through the recombination lines, which represent quantum transitions between high-order excited states of hydrogen atoms. These three types of observation, of continuum emission, continuum absorption and line emission, all involve the square of the electron density, and it will be important to discuss the relation between the required average quantity $\langle n_e \rangle$ and the measured average $\langle n_e^2 \rangle^{1/2}$. The ratio between these depends on the distribution of electrons; it will also appear that the analysis of the observations depends critically on the electron temperature T_e.

Optically there is little possibility of observing the ionised gas. The only direct observation is of faint, diffuse Balmer line emission from the Galaxy, which like the radio recombination lines provides a measure of $\langle n_e^2 \rangle^{1/2}$ (Reay, 1971).

12.2 The galactic HI disc

The neutral hydrogen (HI) interstellar gas is well understood and mapped from measurements of the 21-cm radio spectral line (Kerr, 1969). The well known concentration of HI into spiral arms, which appears in the radio maps, is based on the assumption of strictly circular orbital motion: it is recognised that the non-circular motion which is known to exist could alter the interpretation, so that the spiral arm structure could be a pattern either of velocity components or of density, or a combination of both. We therefore disregard the arm structure, and take a broad average view of the HI distribution as a disc. In this disc the density n_H varies exponentially with distance z from the plane as

$$n_H = n_{H0} \exp(-|z|/z_0). \tag{12.1}$$

The parameters n_{H0} (central density) and z_0 (scale height) vary with galactic radial distance R. According to Kerr the scale height is reasonably constant at 150 pc for $R = 4 \to 10$ kpc, decreasing to 85 pc at the 4 kpc spiral arm, and increasing to 500 pc in the outer arms in the vicinity of $R = 15$ kpc. The central density is about 0.7 cm^{-3} for $R = 7 \to 11$ kpc, decreasing to 0.3 cm^{-3} at $R = 4$ kpc and 0.1 cm^{-3} at $R = 15$ kpc.

The hydrogen gas is expected to be partially ionised; the estimated value of $\langle n_e \rangle = 0.03$ cm^{-3} obtained from the Crab Pulsar suggests that 10% ionisation would adequately account for the pulsar dispersion. The ionisation may be due to low-energy cosmic rays: direct calculation of the expected ionisation is, however, rather difficult since the relevant cosmic

12.2 The galactic HI disc

ray energies are below the range which can be measured directly within the Solar System, so that the flux of ionising cosmic rays is unknown. Another possible source of ionisation is the ultraviolet radiation from supernova explosions. Gerola, Iglesias & Gamba (1973) pointed out that a single supernova explosion might ionise gas over a radial distance of some hundreds of parsecs, and that the lifetime of the ionisation in low-density regions would be long enough for successive supernovae to maintain a high level of ionisation throughout the HI disc. This source would operate preferentially in the inner part of the Galaxy, where it may provide an increased ionisation in the HI disc. Both sources of ionisation will give rise to an increased percentage ionisation in regions of lower gas density, where the recombination rate is smaller. We may therefore expect the scale height for electron density to be greater than that for neutral hydrogen, and indeed we argue later that in the uniform part of the HI disc, extending from about 4 to 10 kpc radius, the electron scale height appears to be at least 300 pc.

The ionised component of the hydrogen disc may be observed most directly through radio surveys of the sky brightness. Maps have been made at frequencies down to 2.1 MHz by Ellis & Hamilton (1966), and measurements from rockets and satellites have extended observations to below 1 MHz, but with poor angular resolution (Smith, 1965: Alexander, Brown, Clark & Stone, 1970). At high galactic latitudes the absorption of extragalactic background radiation reaches optical depth $\tau \sim 1.5$ at 1 MHz, assuming that the absorption is reasonably uniform. For a layer with scale height 300 pc in the distribution of n_e, and correspondingly with a scale height 150 pc in the distribution of n_e^2, and at a temperature T_{1000} in units of 1000 K, this optical depth corresponds to

$$\langle n_e^2 T_{1000}^{-1.5} \rangle = 0.0025. \tag{12.2}$$

The r.m.s. value $\langle n_e^2 \rangle$ would then be 0.05 cm^{-3} for $T = 1000$ K. If $T = 6000$ K, the usual temperature of an HII region, then $\langle n_e^2 \rangle^{1/2} = 0.2$ cm^{-3}. The average $\langle n_e \rangle$ would of course be smaller than $\langle n_e^2 \rangle^{1/2}$ if the distribution were irregular.

The electron disc observed in this way therefore seems to accord well with the expected ionisation of the HI disc, and it also is sufficient to account for pulsar dispersion. The distribution of pulsars through the Galaxy will therefore be discussed in terms of the distances determined from dispersion in this disc; the HII regions play only a minor role, as will be seen in the next section.

The interstellar medium

12.3 The disc of HII regions

From optical work the most prominent sources of interstellar electrons are the ionised clouds known as HII regions. These clouds surround the very hot O and B type stars, and they are therefore concentrated in a thin disc in the inner part of the Galaxy. We consider them in two categories, the large individual clouds and the large number of smaller clouds which may be considered together as a disc.

A classical analysis by Stromgren (1936) showed that the ultraviolet light from a hot star ionises a spherical region whose radius S_0 (pc) depends on the density N (cm^{-3}) of the interstellar hydrogen gas, and on the radius R (in solar radii) and temperature T (K) of the star, according to

$$\log_{10}(S_0 N^{2/3}) = -0.44 - 4.51\frac{5040}{T} + \tfrac{1}{2}\log_{10} T + \tfrac{2}{3}\log_{10} R. \tag{12.3}$$

Within this sphere the ionisation is complete. The density N is somewhat uncertain, and may vary considerably from one HII region to another. It is generally taken to be about 10 cm^{-3}.

An individual HII region near the Sun may contribute significantly to the dispersion measure of a pulsar whose line of sight intersects the region. Prentice & ter Haar (1969) and Grewing & Walmsley (1971) have listed these individual regions and estimated their effect on apparent distances of pulsars.

The averaged effect of the smaller regions was also found by Prentice & ter Haar (1969). A simple approach is indicated by Walmsley & Grewing (1971), who point out that if ρ_i is the space density of stars that would give Stromgren spheres of radius S_i in a gas with density $N = 1$, then in that gas the mean square electron density would be

$$\langle n_e^2 \rangle_1 = \tfrac{4}{3}\pi \sum_i \rho_i S_i^3. \tag{12.4}$$

Equation (12.4) shows that in the practical case where $N \neq 1$,

$$\langle n_e \rangle = \langle n_e^2 \rangle_1 N^{-1} \tag{12.5}$$

so that $\langle n_e \rangle$ can be found from the population statistics of the O and B stars, and an average value of N. Using $N = 14$ cm^{-3}, Prentice & ter Haar obtain $\langle n_e \rangle = 0.004$ cm^{-3}, which is small compared with the electron density in the HI disc. The average contribution of the HII regions can only contribute appreciably to pulsar dispersions if the density N is locally of order 1 cm^{-3} or less.

12.3 The disc of HII regions

Within the disc of HII regions the electrons will be distributed very non-uniformly, so that the dispersion measure (DM) of a pulsar close to the plane may be an unreliable measure of actual distance. The HII regions may also be particularly noticeable though scintillation, the magnitude of which depends on $\langle n_e^2 \rangle$ rather than $\langle n_e \rangle$. We shall later suggest that scintillation effects within the disc are so large that it presents an impenetrable barrier to normal pulsar observations (Chapter 14). It is therefore important to note its geometrical shape, which of course is the same as the population distribution of the O and B stars. These early-type stars are concentrated in the spiral arms, and particularly in the inner spiral arms of the Galaxy. The mean density $\langle n_e \rangle$ will fall with distance $|z|$ from the galactic plane in the same way as the star population, except that the density N may also fall, giving a slower decrease in $\langle n_e \rangle$. Kurochkin (1958) showed that the star density falls exponentially as $\exp(-|z|/z_0)$ where $z_0 = 76$ pc. This represents a thin layer compared with the HI disc, and we should only expect to observe scintillation or dispersion effects due to the HII disc in directions at low galactic latitudes and only in the inner parts of the Galaxy.

Another indication of the relative unimportance of HII regions in this context is given by Davidson & Terzian (1969), who point out that HII regions only occupy a small proportion of interstellar space in the neighbourhood of the Sun. The line of sight to a pulsar in the galactic plane has a 'mean free path' between intersections with HII regions of at least $1\frac{1}{2}$ kpc, which is large compared with the distances of many pulsars.

The disc of HII regions may be observed directly through the hydrogen recombination lines. These are quantum transitions of very high order, producing radio spectral lines with strength proportional to $n_e^2 T^{-1.5}$. These lines cannot unfortunately be observed in the large-scale HI disc, where $\langle n_e^2 \rangle$ is very much smaller. In the HII disc Gottesman & Gordon (1970) observed the line intensities integrated through the disc, and found

$$\langle n_e^2 T_{6000}^{-1.5} \rangle = 1.4 \qquad (12.6)$$

where T_{6000} is the temperature measured in units of 6000 K.

There have been attempts to separate out n_e and T from this result by measuring also the thermal bremsstrahlung radiation from the same direction, which is proportional to $\langle n_e^2 T^{-0.5} \rangle$; it is difficult, however, to separate the thermal and non-thermal components in the radiation, and consistent results have not yet been obtained. Assuming that the temperature is 6000 K, which is typical of HII regions, the Gottesman & Gordon result shows that the mean square electron density is of order unity. Since

The interstellar medium

the distribution is so irregular this is not inconsistent with the estimated mean density $\langle n_e \rangle = 0.004$.

12.4 A model of the electron distribution

Bearing in mind the uncertainties of the foregoing discussions, and in particular the problem of finding $\langle n_e \rangle$ when the most useful measurements provide only $\langle n_e^2 T^{-1.5} \rangle$, it seems reasonable to compare the pulsar observations with the following model:

(i) A thick layer, with density n_e of order 0.03 cm^{-3} at $z = 0$, and scale height 300 pc. The scale height may decrease closer to the galactic centre, and increase beyond $R = 10$ kpc, following the HI thickness.

(ii) A thin layer, comprising the known HII regions and a more diffuse ionisation from other hot stars, concentrated to the plane with scale height 76 pc, and concentrated towards the galactic centre, mainly within a radius of 7 kpc from the centre. The electron density in this layer is very variable, so that irregular refraction and scintillation may be important.

We turn now to the observational evidence provided by the pulsars themselves.

12.5 Absorption in neutral hydrogen

Pulsars at low galactic latitudes may be observed through the spiral arm structure of the Galaxy, and their radio spectra will then show HI absorption at 21-cm wavelength. Apart from small details, the velocity structure of this absorption may be interpreted directly as distance, so that the line profile gives immediately a minimum distance beyond which the pulsar must lie.

The pulsating nature of the source is a great advantage in this measurement. The spectrum of the 21-cm line as observed directly includes the emission spectrum of hydrogen in the telescope beam, and the superposed absorption spectrum of the pulsar is only a small perturbation. The pulsation means, however, that the perturbation is detectable by comparing the spectrum during the pulses with the spectrum between the pulses. A typical observation integrates this difference over some thousands of pulses.

The first HI absorption measurements, made by de Jager, Lyne, Pointon & Ponsonby (1968) were of PSR 0329+54. The absorption spectrum showed the main features of the hydrogen emission spectrum in the same part of the sky, showing that the pulsar lay beyond the major spiral arms. Later observations show that the absorption reaches only to the lower velocity part of the Perseus arm, which covers -25 km s^{-1} to -58 km s^{-1} in the emission spectrum, so that the pulsar must be situated

12.5 Absorption in neutral hydrogen

within that arm at a distance of 2 kpc (A. G. Lyne & R. S. Booth, unpublished work at Jodrell Bank). Fig. 12.1 shows a comparison of the emission and absorption spectra of four pulsars recorded by Graham et al. (1974).

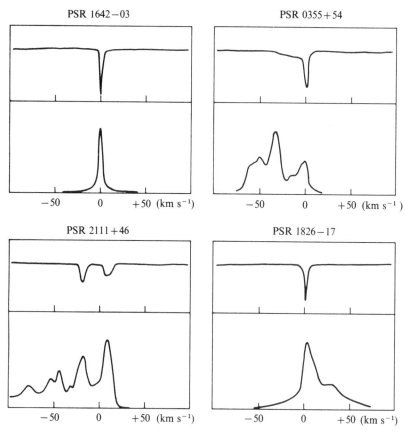

Fig. 12.1. Hydrogen line absorption spectra in pulsars. The top trace shows the absorption spectrum, and the bottom trace the hydrogen emission spectrum. Spectral features which are seen both in absorption and emission correspond to hydrogen gas in front of the pulsar. (After Graham et al., 1974.)

Table 12.1 lists the absorption spectra available in 1975. The distances obtained from the spiral arm model are used to calculate the distance z from the galactic plane.

HI absorption measurements are most valuable for the pulsar distances which they give through direct geometrical interpretation of velocity components, but it is also interesting to measure the depth of the

The interstellar medium

TABLE 12.1 *Pulsar distances determined by H-line absorption*

PSR	l	b	Absorption features (km s^{-1})	Distance (kpc)	z (pc)	References
0138+59	129.1	−2.3	−5.5, −50	∼3	∼120	9
0329+54	145.0	−1.2	−58	2.6	50	7, 10
0355+54	148.1	0.9	−12	∼2	∼30	8, 9
0525+21	183.8	−6.9	0	2	240	8
0540+23	184.4	−3.3	+10	2	120	9
0736−40	254.2	−9.2	0	∼1.5	∼240	7
1642−03	14.1	26.1	+2.7	0.16	65	8, 9
1706−16	5.8	13.7	0	−	−	5
1718−32	354.5	2.5	−4, +9	>1	>45	8
1749−28	1.5	−1.0	0	<1	<20	1, 2
1818−04	25.5	4.7	0	<1.5	<120	6
1826−17	14.6	−3.3	+4, +11	>1.5	>90	9
1929+10	47.4	−3.9	0	<1	<70	5
1933+16	52.4	−2.1	+56	>6	>220	4, 7
1946+35	70.6	5.0	−20.6	>8.5	>850	9
2016+28	68.1	4.0	+14	<1	<70	3, 7
2020+28	68.9	−4.7	+13	<2	<140	8, 9
2021+51	87.9	8.4	+8	<1	<150	7
2111+46	89.0	−1.3	−24, +11	4 → 6	90 → 135	9
2319+60	112.0	−0.6	−53	>2.5	>150	8, 9

1, Guélin, Guibert, Hutchmeier & Weliachaw (1969); 2, Hjellming, Gordon & Gordon (1969); 3, Encrenaz & Guélin (1970); 4, Guélin, Encrenaz & Bonazzola (1971); 5, Gómez-González, Falgarone, Encrenaz & Guélin (1972); 6, Gómez-González, Guélin, Falgarone & Encrenaz (1973); 7, Gordon & Gordon (1973); 8, Gómez-González & Guélin (1974); 9, Graham, Mebold, Hesse, Hills & Wielebinski (1974); 10, A. G. Lyne & R. S. Booth. Unpublished work at Jodrell Bank.

absorption integrated through the profile. If the optical depth at velocity V(km s^{-1}) is τ_v, and the spin temperature of the hydrogen is T_s, then the number of hydrogen atoms per square centimetre of column in the line of sight is

$$N_H = 1.82 \times 10^{18} T_s \int \tau_v \, dV \, \text{cm}^{-2}.$$

The temperature T_s can only be established from the line emission profiles, either by assuming that a peak brightness temperature is close to T_s or by assuming that the widths of the narrow components of the line profiles are determined by T_s and not by turbulent velocities. Assuming that $T_s = 100$ K, Gordon, Gordon & Shalloway (1969) obtained column

12.6 Pulsars at high galactic latitudes

densities of 2.8×10^{21} cm^{-2} for PSR 0329+54 and 0.7×10^{21} cm^{-2} for PSR 1749−28. The ratio of these column densities to the corresponding dispersion measures is found to vary by a factor of ten between different pulsars (Hjellming, Gordon & Gordon, 1969); this is not surprising since the pulsars are close to the galactic plane where the ionisation may be very variable.

12.6 Pulsars at high galactic latitudes

We now interpret the dispersion measures of the pulsars, using the simple model distribution of electrons in the Galaxy. Pulsars at high galactic latitudes will not be affected by the thin disc of HII regions, and we may expect their dispersion measures to refer to the thick disc, plus some effect of individual HII regions from nearby O and B stars.

The dispersion measures (DM) of fifteen pulsars at high latitudes ($|b| > 30°$) are shown in the histogram of Fig. 12.2(a). All but two have $DM < 20$ pc cm^{-3}. Grewing & Walmsley (1971) show that these large values of DM are accounted for by an individual HII region at 170 pc distance. We may therefore describe the distribution of DM as roughly uniform out to a maximum value of $DM = 20$, with a possible concentration at the higher values.

The distribution is consistent with a population of pulsars spreading in $|z|$ to a greater distance than the scale height of the electron gas. The lines

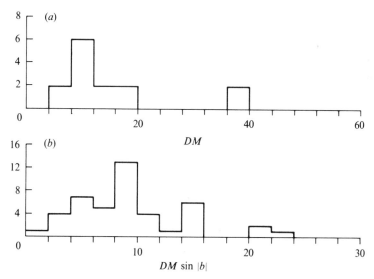

Fig. 12.2. (a) Histogram of dispersion measure (DM) for pulsars at high galactic latitude $|b| > 30°$. (b) Histogram of $DM \sin |b|$ for pulsars with $|b| > 5°$.

The interstellar medium

of sight to these pulsars are inclined at angles of up to 60° from the normal to the galactic plane. Fig. 12.2(b) shows the histogram of $DM \sin|b|$, representing the dispersion measure for a line of sight perpendicular to the galactic plane. Again ignoring some individual high values, the distribution is fairly uniform out to $DM = 12$ pc cm^{-3}, which may be taken as the value for the whole thickness of the disc. If this is equated to the scale height of 300 pc (Section 12.4) the mean electron density on the galactic plane $\langle n_{e0} \rangle$ is 0.04 cm^{-3}. Furthermore, half of the pulsars are seen to be at z-distances of more than 300 pc from the plane, and some may be effectively right outside the HI disc.

Despite the risk of over-interpreting the meagre data provided by these fifteen pulsars, we may attempt to find their distribution in z-distance by assuming an exponential fall of electron density with $|z|$ according to

$$n_e = 0.04 \exp - (|z|/300) \text{ cm}^{-3}.$$

The z-distance of a pulsar with dispersion measure DM at latitude b is then given by

$$z = 300 \log_e \frac{DM \sin|b|}{12} + 1 \text{ pc}.$$

These z-distances for the fifteen pulsars are shown in Fig. 12.3, which again suggests that the pulsars are widely distributed, extending to 500 pc or more from the plane of the Galaxy.

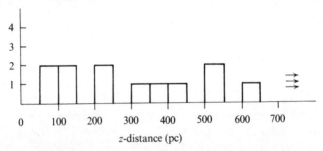

Fig. 12.3. Distribution of fifteen pulsars in z-distance.

This wide distribution in z-distance will be discussed further in Chapter 20, which is concerned with the total population of pulsars in the Galaxy.

References

Alexander, J. K., Brown, L. W., Clark, T. A. & Stone, R. G. (1970). *Astron. Astrophys.* **6**, 476.

Davidson, K. & Terzian, Y. (1969). *Nature, Lond.* **21**, 729

de Jager, G., Lyne, A. G., Pointon, L. & Ponsonby, J. E. B. (1968). *Nature, Lond.* **220**, 128.

References

Ellis, G. R. A. & Hamilton, P. A. (1966). *Astrophys. J.* **146**, 78.
Encrenaz, P. & Guélin, M. (1970). *Nature, Lond.* **227**, 476.
Gerola, H., Iglesias, E. & Gamba, Z. (1973). *Astron. Astrophys.* **24**, 369.
Gómez-González, J., Falgarone, E., Encrenaz, P. & Guélin, M. (1972). *Astrophys. Lett.* **12**, 207.
Gómez-González, J. & Guélin, M. (1974). *Astron. Astrophys.* **32**, 441.
Gómez-González, J., Guélin, M., Falgarone, E. & Encrenaz, P. (1973). *Astrophys. Lett.* **13**, 229.
Gordon, C. P., Gordon, K. J. & Shalloway, A. M. (1969). *Nature, Lond.* **222**, 129.
Gordon, K. J. & Gordon, C. P. (1973). *Astron. Astrophys.* **27**, 119.
Gottesman, S. T. & Gordon, M. A. (1970). *Astrophys. J.* **162**, L93.
Graham, D. A., Mebold, U., Hesse, K. H., Hills, D. Ll. & Wielebinski, R. (1974). *Astron. Astrophys.* **37**, 405.
Grewing, M. & Walmsley, M. (1971). *Astron. Astrophys.* **11**, 65.
Guélin, M., Encrenaz, P. & Bonazzola, S. (1971). *Astron. Astrophys.* **14**, 387.
Guélin, M., Guibert, J., Hutchmeier, W. & Weliachaw, L. (1969). *Nature, Lond.* **221**, 249.
Hjellming, R. M., Gordon, C. P. & Gordon, K. J. (1969). *Astron. Astrophys.* **2**, 202.
Kerr, F. J. (1969). *Ann. Rev. Astron. Astrophys.* **7**, 39.
Kurochkin, N. E. (1958). *Soviet Astron.* **35**, 74.
Prentice, A. J. R. & ter Haar, D. (1969). *Mon. Not. R. astron. Soc.* **146**, 425.
Reay, N. K. (1971). *Mon. Not. R. astron. Soc.* **151**, 299.
Smith, F. G. (1965). *Mon. Not. R. astron. Soc.* **131**, 145.
Stromgren, B. (1936). *Astrophys. J.* **89**, 526.
Walmsley, M. & Grewing, M. (1971). *Astrophys. Lett.* **9**, 185.

13
The interstellar magnetic field

13.1 Eighteen orders of magnitude down

The magnetic fields which are first brought to mind in any discussion of pulsars must surely be the enormous fields at the surface of the neutron star itself. These fields are of the order of 10^{12} gauss. Pulsar radio waves may, however, be used as a means of measurement of the magnetic field on the line of sight to the observer; this interstellar magnetic field is, by way of extreme contrast, of order 1 to 10 microgauss. The observational link between these two extremes is simply stated: the high field of the pulsar is responsible for the polarisation of the radio emission, and the polarisation is used for exploring the interplanetary field through Faraday rotation.

The existence of an interstellar magnetic field was first demonstrated by the observation of the polarisation of starlight, notably by Hall & Mikesell and by Hiltner in 1949. A small degree of linear polarisation was found, which was common to several stars in the same general direction. The polarisation was due to scattering in interstellar space by particles which are elongated, so that they scatter anisotropically, and which are aligned with their long axes perpendicular to the galactic plane. For some time it was thought possible that these particles might be ferromagnetic, so that they would be aligned along the magnetic field, but it is now agreed that they are aligned perpendicular to the field. This occurs through the dissipation of energy in any component of spin which brings the long axis periodically in and out of alignment with the field. Theory suggests that a field of a few microgauss would be sufficient for this process, but no quantitative calculations are possible. The plane of polarisation does, however, provide a very useful measurement of the orientation of the field. The field is evidently well organised over large distances, and we shall later combine the optical and pulsar measurements of field direction to obtain an overall view of the field alignment in the Galaxy.

Radio emission from the Galaxy was already known at the time of the discovery of optical polarisation, but it was not until the synchrotron theory of its origin was developed in the Soviet Union a few years later that the existence of radio emission was used as evidence for a general

13.1 Eighteen orders of magnitude down

magnetic field. The field strength was then estimated to be about 10 microgauss, although this value was derived on the basis of admittedly inadequate information on the flux of cosmic ray electrons. The most significant outcome of the theory was the realisation that the galactic background radiation should be partially linearly polarised. The plane of polarisation should be related to the direction of field, and the degree of polarisation should be related to the smoothness of organisation of the field.

Successful observations of polarisation of the radio background radiation have been made over a range of wavelengths from 21 cm to 2 m. At the longer wavelengths there are severe effects from Faraday rotation between the source and the observer. Faraday rotation is best studied directly by the pulsar observations which are the main subject of this chapter, and we shall not therefore be concerned here with the long wavelength background polarisation. At short wavelengths, where Faraday rotation is small, the degree of polarisation exceeds 10% in several regions of the sky, indicating a well aligned field over a considerable distance. The strongest polarisation is observed at galactic longitude $l = 140°$, where the line of sight appears to be perpendicular to the field over a distance of several hundred parsecs.

A very direct measurement of the magnetic field in some localised regions is provided by the Zeeman effect. The 21-cm radio spectral line of neutral hydrogen is split in a magnetic field, giving two lines of opposite hand of polarisation separated by 2.8 Hz per microgauss. This has been observed in absorption, when the widths of some individual absorption components are small enough to allow the split to be observed. The field strengths so far measured in this way are in the range 10 to 20 microgauss: they refer, of course, to individual absorbing clouds where the density is locally high, and the field strength may not be typical of the interstellar medium.

One other observational method of measuring the interstellar field was available before the discovery of pulsars. Faraday rotation of the plane of polarisation of extragalactic radio sources can be observed for sources over the whole celestial sphere. The rotation is given by

$$\theta = R\lambda^2 \text{ rad m}^{-2} \tag{13.1}$$

where R is the rotation measure of a particular source, and λ is the wavelength of observation. R is determined by the electron density n_e and the component of field along the line of sight, so that

$$R = 0.81 \int Bn_e \cos\theta \, dl \tag{13.2}$$

where B is measured in microgauss, n_e in cm^{-3}, and l in pc.

141

The interstellar magnetic field

The most serious problem in interpreting the observed Faraday rotations of extragalactic sources is in isolating the component which is due to the Galaxy, since there are large rotations intrinsic to some of the sources themselves. A clear pattern of variations of R over the sky does, however, emerge from surveys of many sources. Gardner, Morris & Whiteoak (1969) obtained from this pattern a field directed along the spiral arm, approximately towards longitude $l = 80°$. A more extensive analysis by Vallée & Kronberg (1973), involving 252 sources, essentially confirmed this result.

If the field were parallel to $l = 80°$ at all distances, the pattern of R over the sky would be very simple, with a clear division of sign between two hemispheres. Departures from this simple pattern might be due to rotation intrinsic to the sources; the observed values of R show some large anomalies which are probably due to this. There are, however, some organised patterns which indicate irregularities in the field pattern, to which we will refer later.

The actual strength of the field was not obtainable directly from these Faraday rotations, because the electron density n_e remained unknown.

An account of the various pieces of observational evidence concerning the magnetic field, as they stood in 1968 at the time of the discovery of the pulsars, was given by van de Hulst (1967).

13.2 Faraday rotation in pulsars

The discovery of a high degree of linear polarisation in pulsar radio signals (Lyne & Smith, 1968) led at once to the most direct method of measuring the interstellar magnetic field. Faraday rotation of the plane of polarisation can be measured, as for the extragalactic radio sources, by measuring the position angle of the polarisation at several radio frequencies. In contrast to the extragalactic sources, there is little or no rotation intrinsic to the pulsar, so that the uncertainty in the origin of the rotation is removed. The main improvement over the extragalactic sources is, however, the availability of the dispersion measure DM at the same time as the rotation measure R. This removes most of the uncertainty in interpreting (13.2), since it provides a value for the electron content on the line of sight.

The ratio R/DM, where

$$R = 0.81 \int n_e B \cos \theta \, dl, \quad DM = \int n_e \, dl \qquad (13.3)$$

gives a value of field $\overline{B \cos \theta}$, which is an average over the line of sight. The average is weighted by the electron density along the path. If, for instance, a large part of the dispersion measure in a particular pulsar is

13.3 Configuration of the local field

due to a single HII region, then the field in that region receives a heavy weighting. If a pulsar lies at a large distance above the galactic plane, its rotation measure does not depend much on the magnetic field in its vicinity, but mostly on the field within one scale height of the electron density distribution. The method is obviously difficult to apply if the direction or strength of the field varies considerably along the line of sight; fortunately the scale of variation is large, and the average field $B \cos \theta$ turns out to be a most useful measure, at least of the local galactic magnetic field.

Faraday rotation is obtained from the variation of position angle with frequency. This is usually obtained over a wide frequency range, although at low frequencies where the rotation is large it may be convenient to measure a small change in rotation $\Delta\theta$ between adjacent frequencies ($\nu, \nu + \Delta\nu$ MHz). Then

$$\Delta\theta = -2R\left(\frac{300}{\nu}\right)^2 \frac{\Delta\nu}{\nu}. \tag{13.4}$$

At frequencies near 300 MHz values of $\Delta\nu$ of a few megahertz are appropriate for most rotation measures.

The first measurements of Faraday rotation by Smith (1968) referred to the pulsar PSR 0950+08, which is highly linearly polarised. It has a very small dispersion measure, and its line of sight happens to be nearly transverse to the local magnetic field. Consequently this pulsar has almost the smallest known rotation measure, and the first measurements gave only a small Faraday rotation which could be accounted for by ionospheric rotation alone. The mean interstellar field component was found to be less than 0.2 microgauss; later measurements by Manchester (1972) revised this to 0.7 ± 0.3 microgauss. Lyne, Smith & Graham (1971) obtained field values up to 3 microgauss for several pulsars, and it became clear that the low value for PSR 0950+08 was an unlucky accident.

A series of accurate results by Manchester (1974) now give rotation measures for thirty-eight pulsars distributed over a wide range of directions. Further unpublished results have been obtained by Morris at Jodrell Bank. These results are combined in Table 13.1, and displayed in Fig. 13.1 as average field components.

13.3 The configuration of the local field

Most of the pulsars referred to in Fig. 13.1 are comparatively local, so that the field components are averages over lines of sight within the local spiral arm. These components are shown as circles whose area is proportional to field strength. The filled circles represent positive rotation measure, corresponding to fields directed towards the observer.

The interstellar magnetic field

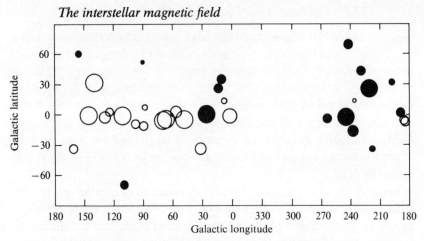

Fig. 13.1. Magnetic field components from Faraday rotation measurements on pulsar radio waves. (After Manchester, 1974.)

There is a remarkably clear-cut division between positive rotation measures, which occur in the 180°–360° hemisphere, and negative rotation measures, which occur in the 0°–180° hemisphere. This indicates immediately that the field is well aligned out to a distance of several hundred parsecs, and that it is directed tangentially, roughly towards $l = 90°$. There are some irregularities, as might be expected from work on the polarisation of background radio emission: for example Wilkinson & Smith (1974) suggested that irregularities exist on a scale of some tens of parsecs, and with a field strength comparable with the mean field. There is also an extensive irregularity in field associated with the North Galactic Spur, which is believed to be the remnant of a supernova shell. The best measure of the mean field is obtained by omitting pulsars more distant than 2 kpc, those that are at high galactic latitudes, and those in the North Galactic Spur. Manchester obtained in this way a local field strength of 2.2 ± 0.4 microgauss, directed toward longitude $l = 94° \pm 11°$.

This field is to be regarded as the average over distances between about 100 pc and 1 kpc from the Sun. The pulsar rotation measures undoubtedly give the best determination of the field in this range of distances, but some considerable deviations in field direction may exist within 1 kpc which are not evident from these measurements alone. The pattern of background polarisation shows that a line of sight towards $l = 140°$ somewhere within 1 kpc intersects the field at right-angles, indicating a field direction towards $l = 50°$ (Bingham & Shakeshaft, 1967; Vallée & Kronberg, 1973). There are four pulsars in this region of sky with known rotation measures, and it is interesting to see if the measures agree with the normal field direction $l = 94°$ or the anomalous direction $l = 50°$.

13.4 The more distant field

TABLE 13.1 *Distances and average interstellar magnetic fields for four pulsars*

PSR	l	b	Distance (pc)	$\overline{B\cos\theta}$ (μG)
0809+74	140	32	230	−2.5
0950+08	229	44	120	+0.7
1133+16	242	69	150	+1.0
1929+10	47	−4	125	−3.3

Table 13.1 gives the distances of four pulsars, obtained simply from their dispersion measures and using an electron density $n_e = 0.025$ cm^{-3}, together with the average line-of-sight field determined by their rotation measures. A local field direction $l = 50°$, $b = 0°$ is incompatible with the measured field towards the first two pulsars; the best value to fit all four is in the region of $l = 90°$, $b = 30°$. Although not much importance should be given to a determination based on so few pulsars, there is a strong suggestion that the region of anomalous field direction must be at a greater distance than 100 pc.

The discovery of more pulsars with very low dispersion measures, if any still exist to be found, would be very valuable in delineating the local field in more detail. Probably the field shows some major loop-like distortion, similar to the North Galactic Spur, which has not yet been properly located. Gardner *et al.* (1969) showed that the apparent field direction, as determined from Faraday rotation in extragalactic sources, differs systematically between groups of sources at different galactic latitudes. This represents an organised distortion of the field which might be a loop; other configurations are, however, possible, such as a helical structure suggested by Mathewson (1968).

13.4 The more distant field

Figure 13.2 shows a plot of average field component against dispersion measure *DM*. The large values of dispersion measure represent large distances, where it is likely that different field directions might be encountered. Although not many points are available for large *DM*, there is a notable lack of large average field values at large *DM*. Using the distance scale discussed in Chapter 12, it is clear that the organised field persists to a distance of at least 1 kpc, and that beyond this distance there are major changes in field direction.

The pulsar PSR 1933+16 is at least 6 kpc from the Sun, as seen from hydrogen absorption measurements. The rotation measure is low, giving

The interstellar magnetic field

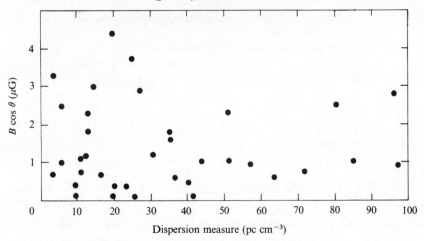

Fig. 13.2. Average field component versus dispersion measure.

a mean field of about 10^{-8} gauss. In contrast, the rotation measure of PSR 1929+10, which is in the same part of the sky but only at 1% of the distance, is four times larger, and the derived average field is over 3 microgauss. Evidently the field reverses at least once over the longer line of sight, giving an average value which is fortuitously low.

13.5 The galactic magnetic field: summary

The pulsar results establish conclusively that the local field is directed along the spiral arm, close to $l = 90°$, and that the field strength is 2 or 3 microgauss. This local field extends to a distance of about 1 kpc; in some directions it extends further, as may be seen by the large rotation measures of some extragalactic sources. For example, 3C410 has $R = -220 \pm 7$ rad m^{-2}, while the nearby pulsar PSR 2020+28, at a distance of about 550 pc, has $R = -44 \pm 2$ rad m^{-2}; the field in this direction ($l = 69°$, $b = -5°$) must therefore extend well beyond 1 kpc distance without much alteration.

On a smaller scale there is a loop distortion associated with the North Galactic Spur. Another distortion which appears as a feature in the distribution of rotation measures of extragalactic sources, and in the background polarisation, may also be a loop or bend in the field, but its distance is unknown. The smaller scale distortions deduced from the background polarisation are not obvious in the pulsar rotations; apparently they do not dominate the field pattern in most directions.

The longitudinal field, directed towards 90° longitude, is to be expected as a result of differential galactic rotation. The irregularities are less obviously explicable, although it is reasonable to suggest that they result from supernova explosions, which can impart sufficient momentum to the interstellar gas for it to drag the field out into a loop.

At larger distances, corresponding to the spiral arms closer to the centre of the Galaxy, the pulsar rotation measures indicate only that the field is no longer closely related to the local field. Optical polarisation results presented by Mathewson & Ford (1970) show that the alignment of polarisation vectors for stars between 2 and 4 kpc distant is better than that for more local stars. The field therefore remains aligned in the plane of the Galaxy, even though the pulsar results indicate that its direction may reverse.

References

Bingham, R. G. & Shakeshaft, J. R. (1967). *Mon. Not. R. astron. Soc.* **136**, 347.
Gardner, F. F., Morris, D. & Whiteoak, J. B. (1969). *Aust. J. Phys.* **22**, 813.
Lyne, A. G. & Smith, F. G. (1968). *Nature, Lond.* **218**, 124.
Lyne, A. G., Smith, F. G. & Graham, D. A. (1971). *Mon. Not. R. astron. Soc.* **153**, 337.
Manchester, R. N. (1972). *Astrophys. J.* **172**, 43.
Manchester, R. N. (1974). *Astrophys. J.* **188**, 637.
Mathewson, D. S. (1968). *Astrophys. J.* **154**, L11.
Mathewson, D. S. & Ford, V. L. (1970). *Mem. R. astron. Soc.* **74**, 139.
Smith, F. G. (1968). *Nature, Lond.* **218**, 325.
Vallée, J. P. & Kronberg, P. P. (1973). *Nature Phys. Sci.* **246**, 49.
van de Hulst, H. C. (1967). *Ann. Rev. Astron. Astrophys.* **5**, 167.
Wilkinson, A. & Smith, F. G. (1974). *Mon. Not. R. astron. Soc.* **167**, 593.

14
Interstellar scintillation

Optical scintillation is familiar as the twinkling of stars, and as the shimmer of distant objects seen through a heat haze. At radio wavelengths scintillation is encountered in many different circumstances, because there are many kinds of radio transmission paths which contain the necessary phase irregularities. The solar corona, for example, contains an irregular outflowing gas, which disturbs radio waves passing through it from distant objects to the Earth. The effects of this may be thought of either as refraction or as diffraction; in more general terms the waves are scattered, giving rise to an angular spread of waves and to fluctuations in wave amplitude.

At the time of discovery of the pulsars, the known examples of radio scintillation gave a rapid fading pattern, not very different from the visible twinkling of stars. The comparatively slow, deep fading of radio signals from the pulsars was an entirely new phenomenon, which was first recognised as a form of scintillation by Lyne & Rickett (1968). The basic analysis of scintillation in terms of random refraction in the interstellar medium was presented by Scheuer (1968). He showed that the fluctuations should have a fairly narrow frequency structure, whose width should depend on the distance of the pulsar. Rickett (1969) obtained experimental proof of this, and showed that fluctuations due to scintillation could be clearly distinguished from the various kinds of fluctuation which are intrinsic to the pulsars. Theory and observation are now both well developed, showing such good agreement that scintillation is now used to investigate the interstellar medium.

14.1 A thin screen model

The simplest model is in practice useful for most of the analysis of pulsar scintillation. In Fig. 14.1 random irregularities of refractive index are found between the source and the observer, but effectively concentrated into a thin screen roughly midway along the propagation path. The irregularities have a typical dimension a, and the screen thickness is D. The average refractive index μ is close to unity, so that a variation $\delta\mu$ extending over a length a changes the phase of a wave by an amount

14.1 Thin screen model

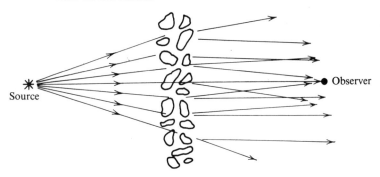

Fig. 14.1. Thin screen model of scintillation.

$\delta\phi \approx (2\pi/\lambda)a\,\delta\mu$. Since for μ close to unity

$$\mu - 1 \approx \frac{n_e e^2}{2\pi m \nu^2} \tag{14.1}$$

where n_e is electron density, the phase change for a wave traversing a single irregularity is

$$\delta\phi = r_e \Delta n_e a \lambda \tag{14.2}$$

where r_e is the classical radius of the electrons (2.82×10^{-13} cm) and Δn_e is the fluctuation in electron density corresponding to $\delta\mu$.

A ray passing through the whole screen encounters D/a irregularities, randomly distributed, so that the difference between the phase perturbations of rays separated laterally by more than a becomes

$$\Delta\phi \approx \left(\frac{D}{a}\right)^{1/2} \delta\phi = D^{1/2} a^{1/2} r_e \Delta n_e \lambda. \tag{14.3}$$

We may now employ either geometric optics to describe the angular spread of rays leaving the screen, or diffraction theory to describe the behaviour of a wave front with phase irregularities $\Delta\phi$ impressed upon it, varying with a lateral scale a. Geometrically, the rays are scattered through an angle

$$\theta_{\text{scat}} \approx \frac{\Delta\phi}{2\pi} \cdot \frac{\lambda}{a} = \frac{1}{2}\left(\frac{D}{a}\right)^{1/2} r_e \Delta n_e \lambda^2 \tag{14.4}$$

and this, under conditions to be discussed later, is the apparent angular size of the source seen through the screen. Separate rays reaching the observer from different parts of the screen will interfere if they have different values of $\Delta\phi$, i.e. if they come from points at least a distance a

apart. Hence if the distance from source to observer is L, and

$$L\theta_{\text{scat}} > a \tag{14.5}$$

a randomly variable amplitude will be observed as the source, or the screen, moves across the line of sight. The ray paths will differ by $\frac{1}{2}\theta_{\text{scat}}^2 L$, which may be a difference of many wavelengths. Interference between rays will therefore differ at different wavelengths; the amplitude will therefore vary over a wavelength difference $\Delta\lambda$ where

$$\frac{\Delta\lambda}{\lambda} \approx \frac{\lambda}{\frac{1}{2}\theta_{\text{scat}}^2 L}. \tag{14.6}$$

In practice the screen is extended along the line of sight so that $L \approx D$. The wavelength range $\Delta\lambda$ is more conveniently expressed as a frequency difference B_s, which for the simple model is given by

$$B_s \approx \frac{8\pi^2 ac}{D^2 (\Delta n_e)^2 r_e^2 \lambda^4}. \tag{14.7}$$

These simple formulae contain the essential features of the scintillation phenomenon as observed. For example, the apparent angular size of pulsars can be measured by interferometry at long wavelengths, when it is confirmed that $\theta_{\text{scat}} \propto \lambda^2$, while Rickett (1969) found $B \propto \lambda^4$ approximately, and B_s decreasing with increasing dispersion measure, as expected if $B_s \propto D^{-1}(\Delta n_e)^{-2}$ and Δn_e was connected with the ionised gas along the line of sight. Section 14.4 gives an account of these observational results.

14.2 Diffraction theory of scintillation

The wave front leaving the screen of Fig. 14.1 may be treated by diffraction theory. A wave front with randomly distributed irregularities of phase $\Delta\phi$ may be constructed from a plane wave with constant phase to which is added a range of other plane waves with a distribution of wave normals, forming an angular spectrum of plane waves (Fig. 14.2). These are the scattered waves, covering an angle of $(2\pi/\lambda)a\Delta\phi$. The amplitude of the waves increases with $\Delta\phi$ until $\Delta\phi$ becomes large compared with 1 radian. The interference of these waves at the observation point is then responsible for the scintillation.

This diffraction analysis is particularly valuable in scintillation theory applying to screens which are extended along the line of sight (Section 14.3). Three simple results emerge, however, which apply equally well to the thin screen model:

14.2 Diffraction theory of scintillation

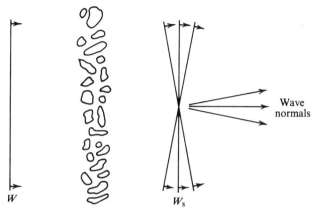

Fig. 14.2. A wave front W, scattered at a diffracting screen, emerges as a range of wave fronts W_s with a distribution of wave normals.

(i) If $\Delta\phi \ll 1$, which is known as weak scattering, there is a large unscattered plane wave. A point source therefore appears to the observer as a point surrounded by a weak scattered halo. As $\Delta\phi$ increases, the point becomes weaker and the halo becomes stronger.

(ii) The lateral scale of the amplitude irregularities in the wave front is the same as the scale a of the phase irregularities provided that $\Delta\phi \ll 1$. When $\Delta\phi \gg 1$, i.e. for strong scattering, the scale is reduced and the typical scale becomes

$$\frac{a}{\Delta\phi} \approx \frac{\lambda}{2\pi\theta_{\text{scat}}}. \tag{14.8}$$

This scale may be observable directly by making simultaneous observations at widely spaced sites, or indirectly by observing the time scale of fluctuations at a single site and assuming that the scintillation pattern moves past the observer with a definite velocity.

(iii) The amplitude of the wave at the observer is found from the addition of the scattered and unscattered waves with random phases. For strong scattering there is no unscattered wave and the amplitude distribution approximates to a Rayleigh distribution. The addition of an unscattered wave changes this to a Rice distribution.

(A Rayleigh distribution of amplitude corresponds to an exponential distribution of intensity. The Rice distribution is discussed in connection with interplanetary scintillation by Cohen, Gundermann, Handebeck & Sharp (1967).)

151

14.3 Thick (extended) scattering screens

The thin screen analysis which has been used so far does not depend critically on the location of the screen provided that it is not very close either to the source or the observer. It is a reasonable deduction that a thick screen will behave similarly to a thin screen provided that neither the source nor the observer is contained within it. A full analysis is nevertheless much more complicated for a thick screen than for a thin screen. Uscinski (1968) analysed the multiple scattering that occurs in successive layers of a thick screen. He showed that the unscattered component decayed exponentially through the screen, so that its relative intensity can be expressed as $\exp(-\beta z)$. Here β is the coefficient of total scattering and z is the distance traversed through the medium; β is given by

$$\beta = \pi^{1/2} r_e^2 (\Delta n_e)^2 a \lambda^2 \tag{14.9}$$

where Δn_e and a are now more precisely defined as follows:

(i) Δn_e is the r.m.s. deviation of electron density; (ii) a is the radial distance within the screen at which the auto-correlation function of Δn_e falls to $1/e$. The irregularities are assumed to have a spherically symmetric Gaussian distribution. (The quantity $(\beta z)^{1/2}$ differs only by a small numerical factor from the less precise quantity $\Delta \phi$ in (14.3).) Weak scattering corresponds to $\beta z \ll 1$, and strong scattering to $\beta z \gg 1$.

The angular spectrum of plane waves emerging from a thickness z in the two extreme cases has a half-width θ_{scat} (to amplitude $1/e$) given by

$$\theta_{\text{scat}} = \frac{\lambda}{\pi a} \quad \text{for} \quad \beta z \ll 1 \tag{14.10}$$

$$\theta_{\text{scat}} = \frac{\lambda}{\pi} (\beta z)^{1/2} \quad \text{for} \quad \beta z \gg 1. \tag{14.11}$$

The value for $\beta z \gg 1$ corresponds to (14.4) above.

Uscinski also shows that full scintillation does not develop until a sufficient distance has been traversed. Corresponding to condition (14.8) above, which ensures that separate rays interfere, he shows that mutual interference between components of the angular spectrum of waves will occur when the distance from the observer to most of the irregularities is large compared with the Fresnel distance z_0 corresponding to the scale size a, i.e. where $z_0 = \pi a^2 / 2\lambda$. Then a sufficient condition for full scintillation to develop is

$$\frac{\beta z}{(\beta z_0)^2} > 6 \quad \text{(approximately)}. \tag{14.12}$$

14.4 Observational results

The close correspondence between the results for thick and thin screens shows that the thin screen analysis is usually sufficiently good for comparison with experimental data; conversely it shows that observational results cannot be expected to distinguish easily between models in which the irregularities are distributed or grouped in various ways between the source and the observer.

14.4 Observational results

Comparisons of theory and observation of pulsar scintillation have been given by Rickett (1970) and by Lang (1971). Scintillation is generally observed as a very deep fading, characterised by a modulation index m close to 1.0. The probability distribution can only partly be separated from the intrinsic fluctuations of the pulsar signal itself; in Fig. 14.3 this is

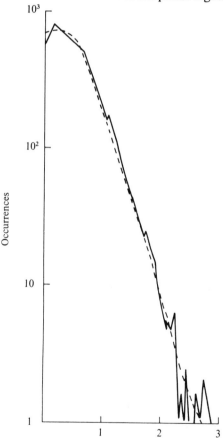

Fig. 14.3. Distribution of pulse intensity at 111 MHz for PSR 1133+16. The broken line is a theoretical distribution based on an exponential law. (After Lang, 1971.)

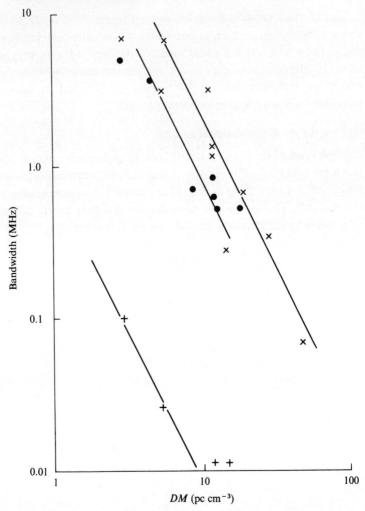

Fig. 14.4. Bandwidth of scintillation (de-correlation frequency B_s) at three observing frequencies: +, 111 MHz; ●, 318 MHz; ×, 408 MHz. The solid lines correspond to theory $B_s \propto (DM)^{-2} \nu^4$.

achieved by smoothing the intensity over a running mean of twenty-five pulses, which unfortunately also reduces the modulation index to $m = 0.7$. Nevertheless the probability distribution is close to the expected exponential distribution.

The speed of fading, which varies inversely with the lateral scale of the scintillation pattern, varies inversely with observing frequency as expected from (14.8).

14.4 Observational results

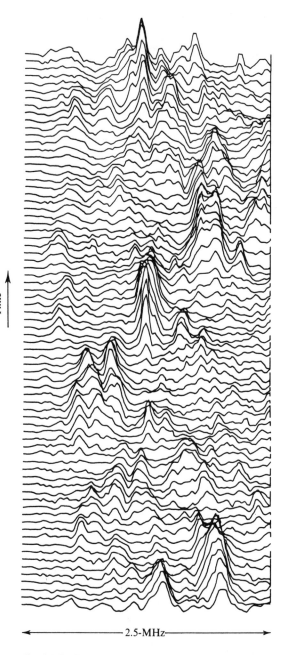

Fig. 14.5. Scintillation of PSR 0329+54. Spectra over a 2.5-MHz band at 408 MHz are displayed at time intervals of 50 s. (After Rickett, 1970.)

Interstellar scintillation

Figure 14.4 shows the dependence of the bandwidth B_s, over which the scintillations are correlated, on the observing frequency ν and the dispersion measure DM for several different pulsars. Equation (14.7) shows that $B_s \propto (DM)^{-2}\nu^4$, corresponding to the solid lines in the figure. As expected, the bandwidth varies as ν^4, but there is a large scatter in the relation with $(DM)^{-2}$, probably due to variations in the distribution of irregularities along the different lines of sight to the various pulsars. The characteristic time scale of scintillation, and the bandwidth B_s, are seen clearly in the series of spectra in Fig. 14.5.

These observations refer to strong scintillations. At higher frequencies the phase variations $\Delta\phi$ may become less than one radian and the scintillation then becomes weak, with a smaller modulation index. This has been observed at 2.4 GHz by Downs & Reichley (1971) for pulsars with small dispersion measures. They find, as expected, that for these pulsars the fading speed becomes independent of frequency above a critical frequency corresponding to $\Delta\phi \approx 1$.

From these observations it is possible to deduce the main parameters of the irregularities in interstellar space with some precision. Rickett (1970) shows that within a factor of 4 the parameters are:

$$\Delta n_e = 4.7 \times 10^{-5} \text{ cm}^{-3}$$

$$a = 10^{11} \text{ cm}.$$

These values are, however, derived for a fairly uniform screen filling the space between source and observer. In Section 14.5 some evidence will be given that the irregularities are more concentrated, in which the value of Δn_e may prove to be somewhat larger.

The scale a is remarkably small on a scale of interstellar distances; it should probably be regarded as a wave motion rather than a cloud-like structure. It should be noted that larger scales of structure would be less effective in causing scintillations, so that a hierarchy of larger sizes might exist, as in typical turbulent situations where large irregularities break down progressively into smaller structures (Scheuer & Tsytovich, 1970). Wentzel (1969) has suggested that the irregularities could originally have been set up by the streaming of cosmic ray electrons through ionised interstellar gas.

14.5 Pulse broadening

Scintillation is the most obvious, but not the only effect of propagation through irregularities in the ionised interstellar gas. Observations of distant pulsars, and particularly at low radio frequencies, which may be

14.5 Pulse broadening

expected to show extreme effects of scintillation, also show the related phenomenon of pulse broadening. Fig. 14.6 shows an example, in which a pulse is drawn out into a long tail at low frequencies. A further example is provided by the Crab Pulsar (Chapter 11), which at frequencies below about 50 MHz shows as a continuous source, in which the pulses are almost completely smeared out by this lengthening process.

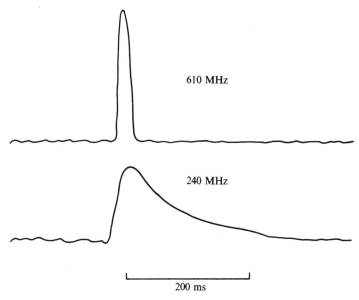

Fig. 14.6. Pulse broadening in PSR 1946+35 due to interstellar scattering. (After Lyne, 1971.)

The close relation between scintillation and pulse broadening is shown by the rough equality

$$2\pi B_s \Delta\tau \approx 1$$

where B_s is the bandwidth of the scintillation, and $\Delta\tau$ is the pulse broadening. This relation, discussed by Lang (1971), is proved experimentally for several pulsars, although in practice it can only be checked by using measurements of B_s and $\Delta\tau$ obtained at widely different frequencies, with an extrapolation of their known frequency dependence to bring them to the same frequency. The relation is easily understood: if ray paths reach the observer over distances differing by $\Delta\tau$, then the relative phases of the various signal components will range over $(2\pi/\lambda)c\Delta\tau = 2\pi\nu\Delta\tau$, so that the relative phases of the interfering waves will change by the order of 1 radian over a band $B_s \approx 1/2\pi\Delta\tau$.

Theoretical treatments of pulse broadening take either the geometrical or diffraction approaches to the problem. A geometric optics approach, involving ray theory, is appropriate for the multiply scattered and greatly broadened pulses, while the diffraction approach may be needed for cases in which multiple scattering is not fully established. We shall consider treatments of these by Williamson (1973) and Uscinski (1974).

14.6 Multiple scattering: geometric approach

Consider first a thin slab of scattering material approximately halfway between the source and the observer, separated by distance L (Fig. 14.7a). If the slab behaves as a thick scatterer, in the sense that it introduces large random phase changes, then emergent rays are scattered with a Gaussian angular distribution, with probability

$$P(\theta) \propto \exp-(\theta/\theta_0)^2.$$

The delay t introduced into a ray deviated by angle θ is $(L/4c)\theta^2$. The probability of a ray deviated between θ and $\theta+d\theta$ reaching the observer is proportional to $\theta\, d\theta$, because of the solid angle subtended by the screen. The probability of a delay $P(t)$ is therefore given by

$$P(t)\, dt \propto \theta \exp-\left(\frac{\theta}{\theta_0}\right)^2 d\theta.$$

Hence the distribution of delay is

$$P(t) \propto \exp-\left(\frac{t}{\tau}\right).$$

A pulse will therefore be observed to have a sharp rise and an exponential decay, with time scale $\tau = L\theta_0^2/4c$. The time scale is therefore proportional to ν^{-4}.

The effect is similar for a screen over a range of positions roughly midway between the source and observer, but if it is close to either source or observer the time scale is shortened proportional to the shorter of the two distances. The theory does not, of course, account for the effects of several screens or of a single physically thick screen.

Extension of the analysis to a scattering medium filling the whole space involves a more complicated ray tracing problem. The ray follows a random walk (Fig. 14.7b) which is constrained to pass through the fixed points of the source and the observer; the problem is to find the distribution of the path lengths. Williamson (1973) has pointed out that this problem is the same as that of Fig. 14.7(c), in which a ray sets out

14.6 Multiple scattering geometric approach

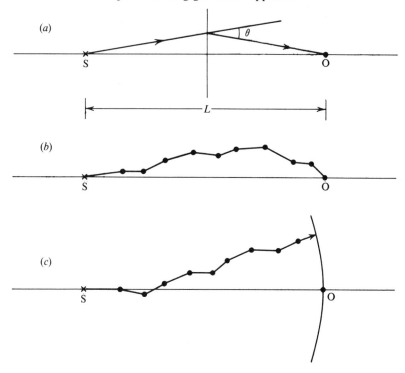

Fig. 14.7. Geometry of multiple scattering. (a) Single scattering; (b) multiple scattering; (c) equivalent geometrical path. S, source; O, observer.

along a fixed direction and ends on a sphere with radius L. The pulse broadening function required is then the distribution of the delay

$$t = \frac{1}{2c}\int_0^L (\theta^2 + \phi^2)\,dz - \frac{1}{2Lc}\left[\int_0^L (\theta\,dz)^2 + \int_0^L (\phi\,dz)^2\right]$$

where θ and ϕ are angular deviations in the x, y planes orthogonal to z. The probability distribution functions of θ and ϕ are both exponentials.

The solution of this integral equation given by Williamson is

$$P(t) = \left(\frac{\pi\tau}{2t^3}\right)^{1/2} \sum_{n\ \text{odd}} \left(\frac{n^2\pi^2\tau}{2t} - 1\right)\exp\left(-\frac{n^2\pi^2\tau}{4t}\right)$$

where the delay time scale τ is given by

$$\tau = \frac{3L\theta_0^2}{2\pi^2 c}.$$

Interstellar scintillation

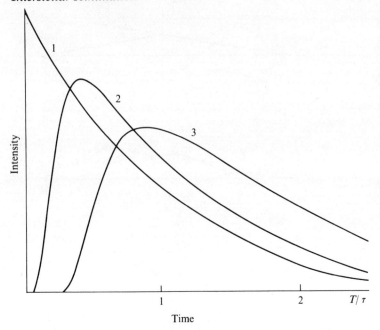

Fig. 14.8. Pulse broadening functions. 1, Scattering from a thin slab; 2, scattering from a more extended region; 3, scattering from irregularities filling the whole line of sight. (After Williamson, 1973.)

This pulse broadening function is shown in Fig. 14.8. For large t it is an exponential with decay time τ, but for small t it rises from zero as

$$P(t) = \left(\frac{\pi^5 \tau^3}{8t^5}\right)^{1/2} \exp\left(-\frac{\pi^2 \tau}{4T}\right).$$

The maximum is at time $t = 0.9\tau$, and there is an appreciable delay at the beginning of the rising section.

14.7 Diffraction theory of pulse broadening

The statistical theory used in the previous section applies to multiple scattering. It is instructive to calculate the effect of low-order, or even single scattering, without the approximation that the scattering is concentrated in a physically thin screen. This requires the diffraction theory developed by Uscinski (1974).

As in his work on scintillation theory (see Section 14.3), Uscinski divides the radiation field into a scattered component $\varepsilon_0 E_r$ and an unscattered component $\varepsilon_0 \gamma(z)$, both being complex quantities. The field in the absence of scattering would be ε_0; scattering increases with

14.7 Diffraction theory of pulse broadening

distance z so that the unscattered field has amplitude reduced by a factor γ such that $\gamma^2 = \exp(-\beta z)$ where β is the coefficient for total scattering. It is important to note that the average of many pulses gives a receiver output which is proportional to $\varepsilon_0^2(E_r^2 + \gamma^2(z))$, i.e. the sum of the powers in the scattered and unscattered components. This is because the relative phase of the two components varies randomly over the receiver bandwidth and also from pulse to pulse. The scattered component, which provides the pulse broadening, is obtained as a detected receiver output

$$P(t) = B \int_{-\infty}^{+\infty} \langle E_r E_s^* \rangle \exp(2\pi i \nu t) \, d\nu$$

where E_r, E_s are two field components within the receiver bandwidth at frequencies separated by an interval ν. B is a normalising constant. The problem is therefore to evaluate $\langle E_r E_s^* \rangle$ in a form suitable for the Fourier transformation of the equation.

The second moment of the field $\langle E_r E_s^* \rangle$ is given by

$$\langle E_r E_s^* \rangle = v(0, 0, \beta z) \exp(-\beta z)$$

where v is the solution of a partial differential equation

$$\frac{dv}{dl} = p(x, y)(1+v) + iA\nabla^2 v$$

in which p is the auto-correlation function of the spatial variation of refractive index, averaged over the z direction, and

$$A = \frac{c(f_r - f_s)}{4\pi \beta f_r f_s}$$

in which β is the total scattering coefficient and f_r, f_s are frequencies of two field components.

Uscinski finds a solution in which there is a series of terms corresponding to the various orders of scattering. Single scattering gives a pulse broadening function

$$P(\tau, \beta z) = \exp(-\beta z) E_1 \left(\frac{\tau}{\beta z} \right)$$

where E_1 is the exponential integral function. The time scale is scaled so that $\tau = 2\pi t/\alpha$, where $\alpha = c/\pi \beta f_0^2 r_0^2$, f_0 being the centre frequency and r_0 the scale of the irregularities (assumed to have a Gaussian distribution).

This first-order scattered component has, as expected, a sharp rise and an exponential fall. All higher order components have a slow rise; the nth order rises as t^{n-1}. All have an exponential fall. The energy in the nth

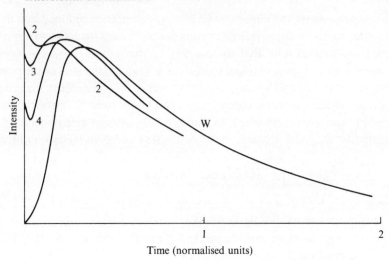

Fig. 14.9. Development of the broadening function for multiple scattering. The theoretical curves correspond to $\beta z = 2, 3,$ and 4. Williamson's curve, corresponding to $\beta z = \infty$, is labelled W. (After Uscinski, 1974.)

component is

$$E_n = \frac{(\beta z)^n \exp(-\beta z)}{n'}.$$

Fig. 14.9 shows the development of the broadening function as βz increases. For $\beta z < 2$ the function is nearly an exponential; the characteristic shape for multiple scattering only develops for $\beta z > 4$.

14.8 Observations of pulse broadening

The Vela Pulsar, PSR 0833−45, has been observed by Ables, Komesaroff & Hamilton (1973) at six frequencies ranging from 150 to 1420 MHz. The pulse shapes (Fig. 14.10) show pulse lengthening with a time scale $\tau_0 = 2.2 \times \nu^{-4}$ ms (ν in MHz). The variation with frequency is as expected. A detailed examination of the pulse shapes by Williamson (1974) provides an interesting comparison with the scattering theory. Fig. 14.11 shows the observed pulse shape at a low frequency (297 MHz) superposed on a theoretical shape derived by convolving the unscattered pulse shape, as seen at high frequencies, with the scattering function for an extended scattering medium localised about either the source or the observer.

Although, as expected, it is difficult to distinguish between the three cases of Fig. 14.8, for this pulsar the third does not provide a fit to

14.8 Observations of pulse broadening

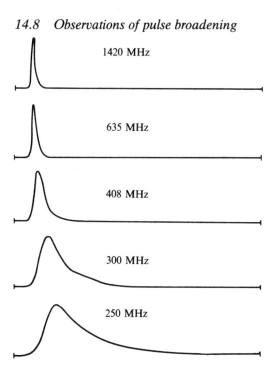

Fig. 14.10. Pulse lengthening in PSR 0833−45. The traces cover the full period of 88 ms. (After Ables *et al.*, 1973.)

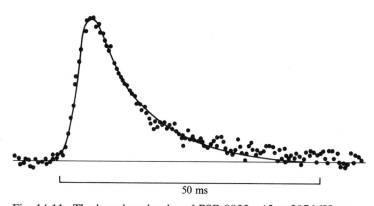

Fig. 14.11. The broadened pulse of PSR 0833−45 at 297 MHz compared with theory for scattering in a region localised around the source or the observer. (After Williamson, 1974.)

observation, and it seems that along the line of sight to this pulsar the irregularities cannot be distributed uniformly.

A clearer indication of the same effect is obtained from pulse broadening in the Crab Pulsar. Here Sutton, Staelin & Price (1971) show that individual strong pulses, which are probably less than 1 ms long at the source, show a very sharp rise followed by an exponential decay with time constant $\tau_0 = 12.2$ ms at 115 MHz. The shape of this broadening is certainly not consistent with the uniform thick scatterer; it indicates instead that the scatterer is concentrated into a small fraction of the line of sight.

In 1974 the pulse broadening of the Crab Pulsar increased dramatically during a period of only a few weeks (Lyne & Thorne, 1975). At 408 MHz the broadening pulse length increased to 4 ms (Fig. 14.12); normally the

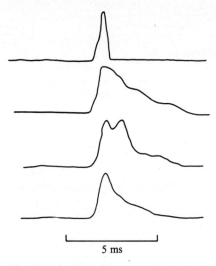

Fig. 14.12. Variable scattering in the Crab Nebula. Recordings at 408 MHz of the interpulse of the Crab Pulsar over a period of 6 weeks. (After Lyne & Thorne, 1975.)

broadening only lengthens the pulse by 50 μs. Again the shape showed a sharp rise followed by an exponential decay, and again the interpretation must be that the scattering occurs in a concentrated region. Variations of the lengthened profile seem to indicate some simple structure within this region, which is probably within the Crab Nebula itself. The movement of the region across the line of sight may be part of the expansion of the nebula, or it may be associated with the active 'wisps' near the pulsar (Chapter 10).

Another indication that the scattering irregularities are not distributed uniformly is the difference between broadening observed in different pulsars with similar dispersion measures. Lyne (1971) pointed out for example that PSR 2002+30, with $DM = 233$, shows no detectable broadening at 408 MHz, while PSR 1946+35 with $DM = 129$, shows a large effect.

Since the irregularities are evidently more or less localised, it might be postulated that they are close to the pulsar. This may be tested by measuring the angular diameters of the pulsars which show pulse lengthening. We now discuss the relation between scintillation and apparent angular diameter, both in theory and observation.

14.9 Apparent source diameters

Scattering in the interstellar medium, which is responsible for scintillation and pulse lengthening, also increases the apparent angular size of the source, as in the familiar examples of optical 'seeing' in telescopes and the radio diffraction in the solar corona. For the simple example of a physically thin screen (Fig. 14.1) the source appears on average as a Gaussian disc with half-width θ_{scat}. If the screen is moved from the halfway position, the apparent angular diameter varies proportional to the distance from the source.

For multiple scattering in a screen filling the whole path Williamson's analysis gives an image diameter proportional to the square root of the path length L. Defining D as the mean square scattering angle per unit path length, the mean square image diameter is $\theta_0^2 = \frac{1}{3}DL$.

Williamson also analyses the cases of physically thick slabs, with multiple scattering, located close to the observer and close to the source. The first gives $\theta_0^2 = DL$, from the definition of D. The second gives $\theta_0^2 = \frac{1}{3}DL(L/\Delta)^2$, where L is the slab thickness and Δ the source distance.

If multiple scattering has not fully developed, diffraction theory may be used to show that there is a component of the wave which is not scattered, providing a virtually point source in the case of a pulsar, together with a scattered wave which represents a scintillation halo round the source. The shape of this halo depends on the scattering properties of the interstellar medium, since it is only for multiple scattering that the angular distribution is necessarily Gaussian.

Observational tests by direct measurement of angular diameters are only possible for multiple scattering, where θ_0 may become large enough for resolution by long baseline interferometry. Resolution has so far only proved possible for the Crab Pulsar, which can be observed at sufficiently low radio frequencies. Cronyn (1970) found an apparent diameter of $2'' \pm 1''$ at 26.3 MHz, and Vandenberg et al. (1973) an apparent diameter of $0.07'' \pm 0.01''$ at 111.5 MHz.

These results may be compared with the theoretical relation between the scale τ_0 of pulse lengthening for a thin screen midway between source and observer

$$\tau_0 = \frac{L\theta_0^2}{4c}.$$

For the Crab Pulsar at 111.5 MHz $\tau_0 = 13$ ms, and $L = 2$ kpc, so that we expect $\theta_0 = 0''.1$. This is sufficiently close to the observed value to exclude any configuration in which the scattering is concentrated solely near the pulsar. (The variable component of lengthening observed for the Crab Pulsar is not necessarily in the same location; this question could, of course, be resolved by a measurement of apparent angular diameter at a time of enhanced pulse lengthening.)

14.10 The velocity of the scintillation pattern

The fluctuations of intensity due to scintillation may represent either a random turbulence in the scattering medium, which changes the configuration of the scintillation pattern over a plane containing the observer, or a relative velocity within the system of observer, medium, and source, so that the pattern drifts past the observer. (The two possibilities are familiar in the patterns of sunlight refracted on to the bottom of a swimming pool, when the waves on the surface are both travelling and changing shape.) Most, but not all, of pulsar scintillation is found to be due to pattern movement rather than pattern instability, so that the rate of the intensity changes is related to the pattern scale and the drift velocity. For example, a typical drift velocity might be 100 km s^{-1}, with a fading time scale of 10 min; the scale of an unchanging pattern would then be about 60 000 km, larger than the diameter of the Earth. On the other hand, the scale could be considerably larger, when the fading time must be largely due to changes in the pattern.

A drifting and unchanging scintillation pattern was observed by Galt & Lyne (1972) for PSR 0329+54. Observations were made at 408 MHz simultaneously at Jodrell Bank and at Penticton, Canada, 6833 km apart. The orientation of this baseline changes as the earth rotates; the pulsar is circumpolar so that a complete rotation of the baseline occurred during 24 hours of observation. The scintillation fluctuations were highly correlated at the two observatories, but with a varying time lag as seen in Fig. 14.13. A sinusoidal variation of this time lag, corresponding to the rotation of the baseline, and with an amplitude of 18 s, is clearly present, corresponding to a velocity of 370 km s^{-1}. This is the velocity of the pattern moving over the baseline; presumably it is due to a large proper motion of the pulsar (Chapter 20). There are also considerable random fluctuations, which may be due partly to pattern instability.

14.10 Velocity of the scintillation pattern

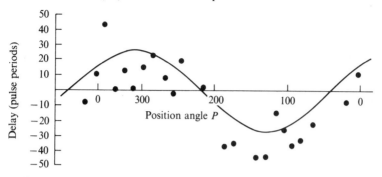

Fig. 14.13. Time lag between fluctuations of PSR 0329+54 recorded at Jodrell Bank and Penticton. The time lag changes sinusoidally as the position angle P changes with the rotation of the Earth. (After Galt & Lyne, 1972.)

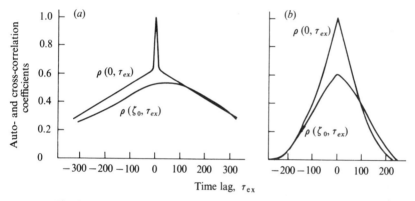

Fig. 14.14. Auto- and cross-correlation functions for scintillation patterns (a) PSR 1929+10 and (b) 1642−03 recorded at separated stations. (After Slee et al., 1974.)

Similar results were obtained by Jones in a single station observation at Jodrell Bank by continuing observations of the fading rate of pulsar scintillation throughout a year. During this time the Earth's velocity through the scintillation pattern varies sinusoidally with amplitude 30 km s^{-1}, so that for an unchanging pattern the time scale of the fluctuations should show a sinusoidal variation; this appears as an annual variation in width of the auto-correlation function of the pattern. Several of the pulsar proper motions discussed in Chapter 20 have been obtained in this way. The values from a single station are, of course, unaffected by assumptions about the lateral scale of the scintillation pattern; they would, however, be affected by instabilities in the pattern if these intrinsic

167

changes were rapid compared with the fluctuations due to pattern drift.

A set of measurements at 326 MHz between Ootacomund (India) and Parkes (Australia) by Slee et al. (1974) showed conclusively that for some pulsars the scintillation pattern was not fully correlated over the baseline of 8000 km, and the cross-correlation of the fluctuations at the two observatories showed an uncertain and variable time lag. In this situation the time lag in the maximum of the cross-correlation tends to be an under-estimation of the delay due to the pattern velocity alone, so that it was for a time thought that the velocities deduced from spaced-receiver experiments might be seriously over-estimated. This situation had already been discussed for the solar wind, as measured by interplanetary scintillation, by Little & Ekers (1971) who in turn referred to classic work by Briggs, Phillips & Shinn (1950) on pattern drifts in ionospheric scintillation.

For a pattern drifting directly along the baseline between the two receivers, and also evolving with time, the analysis proceeds via the general correlation function

$$\rho(\zeta,\tau) = \frac{\{[S(x,t)-\mu_0][S(x+\zeta,t+\tau)-\mu_0]\}}{\{[S(x,t)-\mu_0]^2\}}$$

where μ_0 is the mean intensity. The auto-correlation function at a single station is $\rho(0,\tau)$, and the cross-correlation function for a baseline distance ζ_0 is $\rho(\zeta_0,\tau)$. Examples of these functions as found by Slee et al. (1974) are shown in Fig. 14.14; the cross-correlation function does not reach unity at zero time lag, showing that the fluctuations are not fully correlated. An important result from the analysis by Briggs et al. (1950) can now be applied. The two curves $\rho(0,\tau)$ and $\rho(\zeta_0,\tau)$ are found to intersect at a time lag τ_{ex} where

$$\rho(0,\tau_{ex}) = \rho(\zeta_0,\tau_{ex}).$$

The experimental determination of this time lag τ_{ex} may be used very directly to obtain the pattern velocity V, since

$$\frac{\zeta_0}{2\tau_{ex}} = V.$$

Of course, the pattern will generally be drifting at an angle ϕ to the baseline; this only means that the velocity component $V\cos\phi$ is measured by the delay τ_{ex}. The only assumptions in the analysis are that the intensity pattern is isotropic on the observer's plane, i.e. that the scale size

References

is independent of direction, and that the form of the correlation function is the same for the temporal and spatial co-ordinates.

As might be expected from Fig. 14.14, there is in practice some difficulty in obtaining τ_{ex} accurately. The fading is often rather slow, and statistical fluctuations on the auto- and cross-correlations must be averaged over long periods.

14.11 Proper motions of the pulsars

The relation of these pattern velocities to the actual proper motions of the pulsars depends on the geometrical configuration of the scintillating system. On the simplest hypothesis, that either the source or the observer is moving relative to the scintillating medium, but not both, and that the medium fills the space between source and observer, then the pattern velocity gives directly the velocity of the moving source or observer. If, on the other hand, the scintillation is due to a concentrated screen close to either source or observer, then the pattern velocity is determined largely by the velocity of the nearer of these two points to the screen, through a 'leverage' effect. There is no indication as yet that any pattern velocities are enhanced by such an effect, even though the pulse lengthening observations do suggest that concentrations of scattering irregularities are important.

Finally, there is a possibility that wave motions in the interstellar medium might produce measurable pattern velocities. It is unlikely that organised wave motions would exist over a large enough part of the propagation path for this to happen; even if they did, then the maximum velocities expected are only about 50 km s^{-1}, which is a typical value for magnetohydrodynamic waves in interstellar space.

Since the measured pattern velocities often exceed 100 km s^{-1}, and since there is considerable other evidence for high velocities of pulsars, it may be concluded that the pattern velocities are mainly due to pulsar proper motions and that they provide good measurements of their velocities through the interstellar medium.

References

Ables, J. G., Komesaroff, M. M. & Hamilton, P. A. (1973). *Astrophys. Lett.* **6**, 147.
Briggs, B. H., Phillips, G. J. & Shinn, D. H. (1950). *Proc. Phys. Soc. B* **63**, 106.
Cohen, M. H., Gundermann, E. J., Handebeck, H. E. & Sharp, L. E. (1967). *Astrophys. J.* **147**, 449.
Cronyn, W. M. (1970). *Science*, **168**, 1453.
Downs, G. S. & Reichley, P. E. (1971). *Astrophys. J.* **163**, L11.
Galt, J. A. & Lyne, A. G. (1972). *Mon. Not. R. astron. Soc.* **158**, 281.

Lang, K. R. (1971). *Astrophys. J.* **164**, 249.
Little, L. T. & Ekers, R. D. (1971). *Astron. Astrophys.* **10**, 306.
Lyne, A. G. (1971). *IAU Symposium No. 46*, p. 182. (Dordrecht: D. Reidel.)
Lyne, A. G. & Rickett, B. R. (1968). *Nature, Lond.* **218**, 326.
Lyne, A. G. & Thorne, D. J. (1975). *Mon. Not. R. astron. Soc.* **172**, 97.
Rickett, B. J. (1969). *Nature, Lond.* **221**, 158.
Rickett, B. J. (1970). *Mon. Not. R. astron. Soc.* **150**, 67.
Scheuer, P. A. G. (1968). *Nature, Lond.* **218**, 920.
Scheuer, P. A. G. & Tsytovich, V. N. (1970). *Astrophys. Lett.* **7**, 125.
Slee, O. B., Ables, J. G., Batchelor, R. A., Krishna-Mohan, S., Venugopal, V. R. & Swarup, G. (1974). *Mon. Not. R. astron. Soc.* **167**, 31.
Sutton, J. M., Staelin, D. H. & Price, R. M. (1971). *IAU Symposium, No. 46*, p. 97. (Dordrecht: D. Reidel.)
Uscinski, B. J. (1968). *Phil. Trans. R. Soc. A* **262**, 609.
Uscinski, B. J. (1974). *Proc. R. Soc. A* **336**, 379.
Vandenberg, N. R., Clark, T. A., Erickson, W. C., Resch, G. M., Broderick, J. J., Payne, R. R., Knowles, S. H. & Youmans, A. B. (1973). *Astrophys. J.* **180**, L27.
Wentzel, D. G. (1969). *Astrophys. J.* **156**, L91.
Williamson, I. P. (1973). *Mon. Not. R. astron. Soc.* **163**, 345.
Williamson, I. P. (1974). *Mon. Not. R. astron. Soc.* **166**, 499.

15
Radiation processes

We must admit at the outset of this and the next three chapters that we have a very poor understanding of the processes by which pulsars radiate. We can attempt only to describe the possible processes, and to discuss their possible relevance in the explanation of the observed complex properties of the radiation. Although the energy of the charged particles in the pulsar magnetosphere is derived from an induced electric field, as in a vast electrostatic generator, it is the magnetic field which completely dominates the motion of the particles. The radiation processes operating within the magnetosphere, and giving rise to the pulses of electromagnetic radiation, are therefore due to the acceleration of charged particles by the magnetic field. The processes which may be distinguished as cyclotron, synchrotron, and curvature radiation are extreme cases of this generalisation, but they are sufficiently separate to merit individual analysis. The radiating particles themselves are probably electrons or positrons, although it is possible that ions might also radiate significantly. Generally we will assume that the particles are electrons.

We here present the theory of these radiation processes, leaving a discussion of their operation within pulsars to the next chapter.

15.1 Cyclotron radiation

The radiation from a particle with charge e and mass m, moving with non-relativistic velocity on a circular path in a magnetic field B is at the Larmor frequency

$$\nu_L = \frac{eB}{mc}. \tag{15.1}$$

For an electron $\nu_L = 2.8$ MHz gauss^{-1}.

The radiation is almost isotropic. At angle θ to the axis of the orbit the intensity is

$$I(\theta) = \frac{\pi}{2} \frac{\nu_L^2 \beta^2 e^2}{c}(1+\cos^2\theta) \tag{15.2}$$

where the velocity of the charge is $v = \beta c$.

Radiation processes

The rate of energy loss through radiation is

$$-\frac{dE}{dt} = \frac{8\pi^2}{3} \frac{v_B^2 \beta^2 e^2}{c}. \tag{15.3}$$

For cyclotron radiation at the typical radio frequency of 100 MHz, the loss rate is $2.03 \times 10^{-19} \beta^2$ W.

The polar diagram of the radiation is shown in Fig. 15.1(*a*). The polarisation is generally elliptical, being circular at the two poles and linear at the equator. The Stokes parameters may be normalised to the intensity at the equator, so that for angle θ

$$I = 1 + \cos^2 \theta$$
$$Q = 1 - \cos^2 \theta$$
$$U = 0$$
$$V = 2 \cos \theta. \tag{15.4}$$

When the velocity $v = \beta c$ reaches relativistic values, the gyrofrequency is reduced because of the increased mass of the electron; however, the relativistic velocity also leads to the radiation of harmonics of the gyrofrequency. At high relativistic energies this harmonic radiation

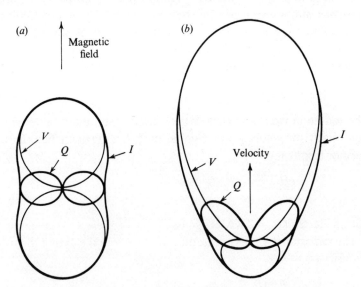

Fig. 15.1. The radiation pattern of cyclotron radiation. (*a*) From an electron in a circular orbit; (*b*) from an electron streaming along the magnetic field. *I, Q, V* are Stokes parameters. (After Epstein, 1973.)

15.2 Cyclotron radiation from streaming electrons

becomes the synchrotron radiation discussed in Section 15.3. But even when most of the radiated power is in the harmonics, the fundamental cyclotron radiation is still emitted with the intensity given by (15.2), with the actual gyrofrequency substituted for ν_L. Although for an individual electron this radiation may now be insignificant in comparison with the harmonic radiation, the coherent radiation from a bunch of electrons may be concentrated in the fundamental and lower harmonics only, and it is therefore important to emphasise that a single highly relativistic electron radiates a fixed power at its frequency of gyration, depending only on the frequency and not on its energy. The frequency is given by

$$\nu_g = \frac{eB}{\gamma mc} \qquad (15.5)$$

where γ is the relativistic factor

$$\gamma = (1 - v^2/c^2)^{-1/2}. \qquad (15.6)$$

The power radiated in the fundamental is given by setting $\beta = 1$ in (15.3). For frequency ν (MHz) the total radiated power in the fundamental is $2.03 \times 10^{-23} \nu^2$ W.

15.2 Cyclotron radiation from streaming electrons

In the pulsar magnetosphere the motion of charged particles must be a combination of gyration about the field lines and streaming along the field lines. For non-relativistic particles the gyroradiation is essentially unchanged by a streaming motion, apart from a Doppler shift in frequency. The more relevant case of highly relativistic particles is usually complex, but a simple case exists for relativistic particles streaming with very small pitch angles (Epstein, 1973).

The pitch angle ψ, which is the angle between the velocity vector and the field vector, must be small compared with γ^{-1} for this simple case. The particle motion may then be considered as a pure gyration in a moving frame of reference, and in this moving frame the motion is non-relativistic. The radiation is therefore monochromatic gyrofrequency radiation, which is observed with a Doppler frequency shift in the stationary frame. The polar diagram and the polarisation are modified as shown in Fig. 15.1(b). The relativistic motion along the field lines beams the radiation in the forward direction, emphasising one circular component.

The more general case for streaming relativistic electrons must be considered as part of the theory of synchrotron radiation.

Radiation processes

15.3 Synchrotron radiation

A charged particle gyrating with high velocity ($\beta \approx 1$) radiates a spectrum of harmonics which extends to frequencies of order $\gamma^2 \nu_L$, i.e. to γ^3 times the gyrofrequency ν_g. When γ is large, this radiation may be regarded as a continuous spectrum. This radiation is synchrotron radiation, also known as magnetobremsstrahlung.

The theory of synchrotron radiation has been presented in detail by Ginzburg & Syrovatskii (1965, 1969). A simple understanding of the essential points may be obtained by considering the electric field radiated by a single electron, gyrating perpendicular to the magnetic field, and observed in the plane of its orbit. This field consists of pulses each occurring as the electron travels towards the observer. The relativistic velocity of the electron concentrates the field in the forward direction, into an angle γ^{-1}. The observer therefore sees a pulse while the electron traverses an arc γ^{-1} of its orbit, which occurs in time $\gamma^{-1}\nu_g^{-1} = \nu_L^{-1}$. Since the electron is travelling towards the observer at this time the duration of the pulse is further compressed by a factor $(1-\beta)$, with the overall result that the pulse width is approximately $\gamma^{-2}\nu_L^{-1} = \gamma^{-3}\nu_g$. Harmonics of the gyrofrequency are therefore emitted up to a factor γ^3, or up to γ^2 of the Larmor frequency. The full theory defines a 'critical frequency', which is close to this high harmonic frequency.

The radiation is concentrated in the plane of gyration, within an angle γ^{-1} approximately. (The low harmonics are less concentrated in angle; the nth harmonic is concentrated roughly within an angle $n^{-1/3}$ radian.) For particles streaming along a magnetic field line, with pitch angle θ, the radiation is concentrated in a cone with half-angle θ, i.e. the radiation is concentrated in the direction of the electron velocity. The radiation is then determined by the component of the field perpendicular to the electron velocity, so that we may write B in formulae for critical frequency and intensity.

The polarisation of the radiation may be understood from a description of the electron acceleration as seen by the observer. For the simplest case of the observer in the plane of gyration, with no streaming motion, the polarisation is plane (Fig. 15.2a). An observer above or below the plane will see a circular component (Fig. 15.2b, c), giving elliptical radiation, whose ellipticity diminishes to zero at edges of the beam, where the polarisation approaches pure circular. A cross-section across the cone of radiation from an electron with a large component of velocity along the field line shows similar polarisation characteristics.

The spectrum of synchrotron radiation, expressed as the total power radiated in all directions, is shown in Fig. 15.3. Below the frequency ν_m,

15.3 Synchrotron radiation

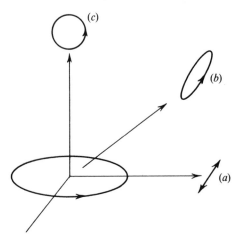

Fig. 15.2. Polarisation of radiation from an electron in a circular orbit. For explanation see text.

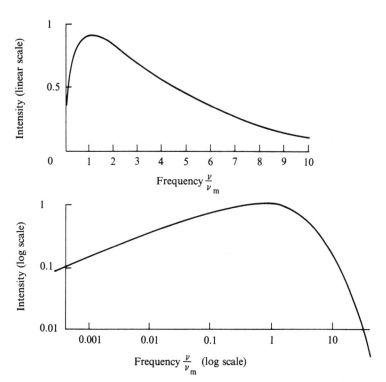

Fig. 15.3. Spectrum of synchrotron radiation, on linear and logarithmic scales.

where the power is a maximum, the spectrum is a power law, proportional to $\nu^{1/3}$; above ν_m it falls exponentially as $\exp[-(\nu/\nu_c)]$.

The essential characteristics of the radiated power are as follows:

Frequency (MHz) in field B (gauss)

$$\begin{aligned}\nu_m &= 1.2 B_\perp \gamma^2 \\ &= 1.8 \times 10^{12} B_\perp (E_{erg})^2 \\ &= 4.6 B_\perp (E_{MEv})^2.\end{aligned} \qquad (15.7)$$

Total radiated power (spectral density) at ν_m

$$p(\nu_m) = 2.16 \times 10^{-22} B_\perp \text{ erg s}^{-1} \text{ Hz}^{-1}. \qquad (15.8)$$

Intensity at frequency ν (approximately)

$$p(\nu) \approx 4 \times 10^{-22} B_\perp \left(\frac{\nu}{\nu_m}\right)^{1/3} \exp\left(-\frac{\nu}{\nu_c}\right) \text{ erg s}^{-1} \text{ Hz}^{-1} \text{ sr}^{-1}. \qquad (15.9)$$

Lifetime of electrons (time to lose half energy by synchrotron radiation)

$$T_s = \frac{5.1 \times 10^8}{\gamma B_\perp^2} \text{ sec.} \qquad (15.10)$$

The polarisation at frequencies near ν_m is shown in Fig. 15.4, which represents a cross-section through the fan beam of radiation from a

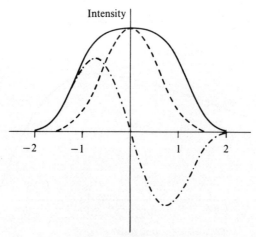

Fig. 15.4. Angular distribution of synchrotron radiation from a collimated monoenergetic electric beam. ———, total power (I); – – –, circular component (V); – · · – · · –, linear component $(Q^2 + U^2)^{1/2}$.

15.4 Curvature radiation

particle moving in an orbit perpendicular to the magnetic field. The polarisation is presented as a linear component $(Q^2+U^2)^{1/2}$ and a circular component V, where I, Q, U, V are the Stokes parameters. The beamwidth is approximately γ^{-1} at frequency ν_m and above; it increases at lower frequencies where it equals approximately $\gamma^{-1}(\nu_c/\nu)^{1/3}$.

It is important to note that the circular polarisation changes hand across the beam. Accordingly, if an ensemble of electrons with a range of pitch angles exceeding γ^{-1} contributes to the radiation, so that the beamwidth of their combined radiation exceeds the width of the beam in Fig. 15.4, the average value of the circular component will be diminished. In practical astrophysical situations this may remove the circular component entirely, leaving only a linear polarisation amounting to about 70% at most.

15.4 Curvature radiation

In the very high magnetic field of the pulsar magnetosphere, an electron may follow the path of a magnetic field line very closely, with pitch angle nearly zero. The field line will generally be curved, so that the electron will be accelerated transversely and will radiate. The radiation, which is closely related to synchrotron radiation, is called curvature radiation.

An electron with relativistic velocity $v \approx c$, constrained to follow a path with radius of curvature ρ, radiates in a similar way to an electron in a circular orbit with gyrofrequency $c/2\pi\rho$. As in synchrotron radiation, there is a critical frequency given approximately by

$$\nu_c \approx \frac{c}{2\pi\rho}\gamma^3. \tag{15.11}$$

The maximum intensity from an electron with energy E_e (eV) is at a frequency $\nu_m \sim 10^{-8} E_e^3 \rho^{-1}$. The spectrum is of the same form as synchrotron radiation, varying as $\nu^{1/3}$ at low frequencies and falling exponentially above ν_c. The intensity depends on γ instead of B; at frequencies below ν_c it is given by Jackson (1962) as:

$$I(\nu) \sim \frac{1}{2\pi}\frac{e^2}{c}\left(\frac{c}{\rho}\right)\gamma\left(\frac{\nu}{\nu_c}\right)^{1/3}. \tag{15.12}$$

The total radiated power is $10^{-38} E_e^4 \rho^{-2}$ W.

As a numerical example we follow the suggestion by Komesaroff (1970) that pulsar radio emission is due to curvature radiation from particles streaming along polar field lines. Then c/ρ is of the order of the pulsar period (≈ 1 s) and ν_c must be at high radio frequencies (≈ 1 GHz) giving $\gamma \approx 10^3$. For these values of curvature and particle energy the intensity from a single electron at frequencies near ν_c is of order 10^{-27} erg s^{-1} Hz^{-1}. This is very much smaller than the cyclotron or synchrotron radiated intensity; however, the intensity is in practice determined mainly by coherence conditions, and intensity alone does not provide sufficient grounds for a choice of radiation mechanism. It is, in any case, possible that much higher values of γ are involved in the curvature radiation.

The polarisation of curvature radiation is essentially the same as that of synchrotron radiation.

15.5 Coherence

The very high brightness temperatures observed in the radio pulses can only be obtained by the coherent motion of charged particles. This coherent motion may either be in the source, or in a maser amplifying region outside the source. We consider arguments for these two possibilities in the next chapter, and here discuss only the effect of coherence within the source.

Goldreich & Keeley (1971) have investigated the stability of a simple system of charged monoenergetic particles constrained to move in a ring. By evaluating the interactions between the particles they show that a perturbation in density may tend to grow. If the scale of this perturbation, expressed as an angular width δ, is small compared with γ^{-3}, then the radiation from the bunch of particles will be in phase over the whole of the synchrotron spectrum, giving enhanced radiation. If there are N particles in the bunch, the radiated intensity will then be proportional to N^2 rather than to N. Larger angular scales of perturbation would provide coherence only for the lower harmonics of the synchrotron radiation; in the extreme a perturbation consisting of a single sinusoidal distribution round the ring would only enhance the fundamental gyrofrequency radiation. This latter possibility is favoured in the discussion in Chapter 18.

As we have seen in the previous section, curvature radiation is very similar to synchrotron radiation, so that coherence has a similar effect. If the radio emission is the curvature radiation from electrons constrained to follow field lines from the polar cap, which have a large radius of curvature, the radiation must be at a high harmonic frequency, with a correspondingly small scale of bunching.

Coherence can only be achieved for isotropic, or nearly isotropic, radiation if the bunches of particles have a scale of less than half a wavelength, as measured at the source of the radiation. If the radiation is beamed, then a disc-like bunch is an effective radiator, provided that the axis of the disc is directed parallel to the radiation. Along the axis of the disc the thickness must be less than half a wavelength, as before; but the width may be very much greater. For an infinite flat disc the coherence effectively operates over one Fresnel half-period zone, whose size depends only on the distance of the observer. More realistically, the disc may have a curvature comparable to the pulsar radius R, giving a zone radius $\sqrt{(R\lambda)}$ for coherence.

The formation of bunches and discs of coherently radiating electrons cannot as yet be explained theoretically.

15.6 Maser amplification

Ginzburg & Zhelesnyakov suggested that the high intensity of pulsar radiation might be due to an amplification outside the source, rather than to a coherence within the source. There are two possibilities for such a 'maser' amplification; a simple negative absorption and a plasma wave coupling. Both derive energy from a non-thermal electron energy spectrum, in which the electrons are divided into highly relativistic streaming particles and a relatively cold plasma. The first involves only a suitable energy spectrum, which allows energy to be transferred to the electromagnetic wave through a coupling between low- and high-energy electrons. The second involves a plasma wave which can grow, again due to the non-thermal energy spectrum. This wave grows as in the classic two-stream instability. The radiation is now amplified by a coupling with the plasma wave.

The difficulty with theories involving maser amplification is to obtain sufficient gain in the simple reabsorption, and to obtain gain over a wide bandwidth in the plasma process. No theory of this kind has yet attempted to explain the detailed characteristics of polarisation in the sub-pulses.

References

Epstein, R. I. (1973). *Astrophys. J.* **183**, 593.
Ginzburg, V. L. & Syrovatskii, S. I. (1965). *Ann. Rev. Astron. Astrophys.* **3**, 297.
Ginzburg, V. L. & Syrovatskii, S. I. (1969). *Ann. Rev. Astron. Astrophys.* **7**, 375.
Ginzburg, V. L. & Zhelesnyakov, V. V. (1970). *Comments Astrophys. Space Phys.* **2**, 197.
Goldreich, P. & Keeley, D. A. (1971). *Astrophys. J.* **170**, 463.
Jackson, J. D. (1962). *Classical Electrodynamics*, p. 467. (New York: John Wiley.)
Komesaroff, M. M. (1970). *Nature, Lond.* **225**, 612.

16
The emission mechanism
I: analysis of observed properties

Before we embark on the difficult task of trying to understand the emission mechanism in pulsars, we should recapitulate the wealth of observational data, listing the properties which seem to be relevant to the physics of the emission process, and seeking correlations between them. We do not yet know the location of the radio emitter, and we do not yet know how to describe the charged particle motions within it. Two approaches can be made to these questions, one through the observed radiation, the other through analysis of the electrodynamics of the pulsar and its magnetosphere. These two approaches do not yet meet, and we must examine the evidence provided by both. We shall first recapitulate the properties of the radio pulses, and especially their width, intensity and polarisation. These properties should lead to a description of the emitter itself. We then turn to the distributions of period P and its derivative $\dot{P}$, seeking any organisation of pulsars according to period P, age $P/\dot{P}$, or slowdown torque $P\dot{P}$. These should lead to an understanding of the magnetosphere, and to the plasma physics responsible for the emission.

16.1 The integrated pulse profiles

The integrated profiles (see Chapter 8) are obtained by adding together some hundreds of pulses. Their shapes known in detail for many pulsars, and they have the following properties:

(1) The angular width of the integrated profile, which is typically 10° of rotation, does not depend on the period.

For most pulsars:

(2) The angular width of the integrated profile is almost independent of radio frequency; where there is a marked dependence it is usually such that the width increases at lower frequencies.

(*Comment*: There is good reason for regarding the width of the integrated profile as delineating an emitting region, within which sources of radiation may occur, rather than as a beamwidth related to an emission process. We shall consider whether this emitting region can be identified with a region of the magnetosphere such as the polar cap or a particular configuration of magnetic field lines near the velocity-of-light cylinder.)

16.2 Individual pulses: the sub-pulses

(3) The shapes of the profiles are stable and repeatable over time intervals ranging from minutes to years.

(4) Two different profiles may be observed from a single pulsar, occurring at random times with characteristic intervals between several seconds and several hours.

(*Comment*: This 'moding' behaviour is usually considered to be a change of excitation across an otherwise stable, defined emitting region.)

(5) The integrated pulse shapes may show several components, often arranged roughly symmetrically.

(6) 'Interpulse' components sometimes occur well away from the main pulse.

(7) Integrated polarisation, found by adding the Stokes parameters of many pulses, is typically mainly linear, with position angle sweeping smoothly and monotonically over 1 or 2 radians.

(8) Integrated polarisation frequently also contains circular polarisation, usually near the centre of the profile, and often changing hand.

(*Comment*: These stable polarisation patterns are usually considered to be related directly to the magnetic field direction at different points across the emitting region.)

Most of the properties of the integrated profiles apply also to the visible and X-ray radiation from the Crab Pulsar. It is particularly remarkable that:

(9) The integrated profile for the Crab Pulsar is very similar over the whole observed range of 40 octaves.

16.2 Individual pulses: the sub-pulses

Wherever measurement of individual pulses (see Chapter 9) has been possible, it has been found that:

(10) Individual radio pulses are very variable in intensity, and

(11) variable in time of arrival, within a range covered by the integrated profile.

These irregularities are more or less evident in the various pulsars. Intensity variations may be described by a histogram of pulse intensities, which is characteristic for individual pulsars. The histogram may show:

(12) Missing pulses, which occur at random times typically between a few periods and a few hundred periods.

(*Comment*: The missing pulses may be regarded as another 'mode' of emission, in which the excitation is zero over the whole emitting zone instead of over only part of it.)

Individual pulses consist of one or more distinct sub-pulses. For those

few pulsars in which these can be observed over a wide range of frequency:
- (13) The width of the individual sub-pulses is practically independent of radio frequency.

(*Comment*: This is a most remarkable observation, since an individual sub-pulse is often regarded as a basic beam of radiation, whose width might be expected to vary with wavelength.)
- (14) Individual sub-pulses are typically highly polarised, often approaching 100%. The polarisation is usually elliptical, i.e. there is a substantial circular component.
- (15) The polarisation changes smoothly through a sub-pulse, so that the plane component may swing over 1 or 2 radians and the circular component may change hand once during the sub-pulse.

(*Comment*: This pattern of polarisation suggests that during the sub-pulse the source of emission is being viewed from a range of angles, so that changes in the aspect of the source are reflected in the changes of polarisation.)
- (16) The visible and X-ray radiation from the Crab Pulsar does not show the same sub-pulse structure: it therefore differs from the radio radiation in that each pulse has the shape of the integrated profile.
- (17) A sub-pulse may appear as a recognisable entity for several rotations of the pulsar. During this time its position within the profile may drift, usually to an earlier phase.
- (18) Fine structure (the 'microstructure') is also seen within a sub-pulse, on a time scale as short as $10\,\mu$s in one pulsar.

16.3 Intensity and spectrum

- (19) The very high intensity of the radio pulses requires the source to be coherent.
- (20) The optical and visible radiation from the Crab Pulsar need not, however, be from a coherent source.
- (21) The radio spectrum obtained by integrating the whole pulse power over many pulses generally follows a power law $I \propto \nu^{+\alpha}$, where the spectral index α is usually about -2 or -3. There is an indication that the spectrum steepens or even cuts off at a high radio frequency, of the order of 3 GHz.
- (22) A definite cut-off is observed in the spectrum of several pulsars below about 100 MHz: other pulsars can be observed down to 10 MHz without any signal of a cut-off.

(*Comment*: The radio spectrum is interpreted by some authors as a direct

16.4 The period P and its derivative $\dot{P}$

indication of broad-band radiation from a single source. There is, however, good reason for suggesting that the source itself is not very broad-band, while the spectrum is reflecting a distribution of more or less narrow-band sources. The changes in integrated profile with frequency may then reflect spatial distribution of sources which radiate predominantly at different frequencies. Further evidence for narrow bandwidth is found in individual pulses (see below).)

(23) The optical and X-ray spectrum of the Crab Pulsar forms a smooth continuum which is distinct and separate from the radio spectrum.

(*Comment*: There is no reason to suggest that this radiation is anything other than broad-band radiation, such as synchrotron radiation.)

(24) The infrared spectrum of the Crab Pulsar shows a fall which is often interpreted as self-absorption.

The spectrum of individual radio pulses is not yet properly investigated. An individual sub-pulse can certainly be detected over a wide frequency range, but there is also evidence for fine frequency structure within this range. Tentatively:

(25) An individual pulse has closely related properties of sub-pulse structure, intensity, and polarisation over a wide range of frequency, inside which its spectrum resembles the integrated spectrum.

(26) Fine frequency structure, over a range of some kilohertz, is associated with the 'microstructure' of pulse intensity.

(*Comment*: It is not clear whether the fine frequency structure is caused by rapid variations in excitation of an inherently broad-band source, or whether it represents separate narrow-band sources of radiation.)

16.4 The period P and its derivative $\dot{P}$

We consider separately (in Chapter 20) the question of pulsar ages and evolutions. In this section we are concerned with any possible relationships between the periods and the radiation process.

The derivative $\dot{P}$ is now known for over eighty pulsars, constituting a group in which there should be no selection effects. The results presented in Chapter 20 show:

(27) The period P ranges over two decades, most periods lying within a range of only three to one.

(28) The derivative $\dot{P}$ ranges over four decades, and it is fairly evenly spread over this range.

(29) There is no apparent correlation between P and $\dot{P}$.

The emission mechanism: I

(*Comment*: These observations seem to rule out any idea that pulsars with different periods form an evolutionary sequence.)

If the periods are lengthening due to the emission of dipole magnetic radiation, so that $\dot\Omega \propto (M^2/I)\Omega^3$, where M is the magnetic moment and I the moment of inertia, then $P\dot P \propto M^2/I$. We therefore examine the range of $P\dot P$ and deduce:

(30) The ratio M^2/I is spread over more than four decades.

(*Comment*: Neutron star models show that I is largely independent of the mass, since the more massive neutron stars have a smaller radius. The spread in M^2/I may, therefore, be regarded as a spread over about two orders of magnitude in dipole moment.)

16.5 Radiation characteristics related to age and period

The rotation rate of a pulsar must influence in some way the rate of energy flow through the pulsar magnetosphere. It certainly determines the radius of the velocity-of-light cylinder, and in simple dipole models it determines the size of the polar cap. We might therefore expect to see considerable differences in the radiation characteristics of slow and fast pulsars. Apart from the existence of light and X-rays from the fastest pulsar, and from no other, the only clear effect is the absence of pulsars with long periods, showing that the radio radiation 'turns off' at a period of about 1 s (but see Section 20.8, Chapter 20).

Some of the more subtle radio characteristics may, however, be related to age. Pulsars which show drifting sub-pulses are generally those with large period and small $\dot P$. We noted in Chapter 9 that the 'negative drifters', in which the sub-pulses drift earlier, are pulsars with smaller $\dot P$ than the pulsars which are 'positive drifters'.

Moding, and missing pulses, are generally observed only in long-period pulsars. This may be regarded as a characteristic of pulsars that are approaching the end of their radiating lives, although there is no evidence that this is so.

It has been suggested that multiple components in the integrated profile are characteristic of longer period pulsars. This is not sufficiently clear-cut characteristic for any account to be taken of it in theoretical work. Similarly, it may possibly be significant that pronounced microstructure is only seen frequently in the short-period pulsar PSR 0950+08, while similar behaviour occurs in the giant pulses of the Crab Pulsar. It cannot reasonably be held, however, that this is a characteristic of short-period pulsars.

No relationship exists between spectrum and period, or between luminosity and period. A possible relation between luminosity and the

16.5 Radiation characteristics

existence of a low-frequency cut-off (Chapter 8) is weak and may be totally insignificant.

We conclude:

(31) The radiation characteristics are not markedly dependent on period or its derivative $\dot{P}$, except that there must be a catastrophic stop to the radiation when the period of a pulsar reaches about 1 s, the actual limit depending slightly on $\dot{P}$.

17
The emission mechanism
II: geometrical considerations

The location of the source of radiation within the pulsar magnetosphere is not at all obvious. The observations of pulse width, pulse shape, and of the sweep of polarisation position angle, provide a wealth of information about the beam of radiation as it crosses the observer's line of sight; we shall explore the geometry of this beam to see how it relates to this question of source location. It might be thought that the whole of the pulse phenomenon could be described as a radiation polar diagram of the emitting source, sweeping across the observer as the neutron star rotates. The description of the actual pulses, as categorised in the previous chapter, suggests that only the sub-pulses can be considered in this way. The integrated pulse profile will in contrast be considered as a longitude distribution of sources, which may of course be broadened by convolution with the sub-pulse angular width. The fine time structure known as microstructure will be taken to be a rapid fluctuation in excitation of the emission, and not therefore the concern of this chapter.

Our main concern will be to use the interpretations of the longitude distribution of sources, and the formation of the basic beam which is seen as a sub-pulse, to find the location of the source. According to various theories this may be close to the surface of the neutron star, or at various distances from it up to, or even beyond, the velocity-of-light cylinder.

17.1 The width of the integrated profile

The angular extent of the distribution of sources, as measured by the width of the integrated profile, must be determined in some way by the configuration of the magnetic field. The family likeness of the various integrated profiles suggests that this particularity in the field cannot be a local irregularity like a sunspot, but must be a general property of a basically dipole field. A simple dipole field does not itself have an appropriate small and well defined region, but a magnetosphere deriving from a rotating dipole field does have several possibilities. The most obvious candidate is the restricted region near the magnetic poles through which flow the open field lines. Fig. 17.1 shows this polar cap region for an aligned dipole.

17.1 Width of the integrated profile

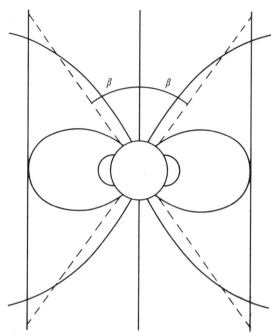

Fig. 17.1. The polar cap is defined by the magnetic field lines which are tangential to the velocity-of-light cylinder. In practical configurations the angular width 2β is of order 1° only.

At the boundary of the polar cap there are closed field lines which just touch the velocity-of-light cylinder. Assuming a simple dipole field throughout, and an aligned dipole, we may find the angular radius β of the polar cap by tracing these lines back to the surface. A field line at angle δ to the dipole axis, at radius r, is at angle δ_s where it cuts the surface at radius a, given by

$$\left(\frac{\sin \delta}{\sin \delta_s}\right)^2 = \frac{r}{a}. \tag{17.1}$$

The field line tangential to the velocity-of-light cylinder, at $r = c/\Omega$, therefore has

$$\sin \delta_s = \left(\frac{\Omega a}{c}\right)^{1/2} = \left(\frac{2\pi a}{Pc}\right)^{1/2}. \tag{17.2}$$

At the edge of the polar cap the line of force is inclined to the radius vector by $\tan^{-1}(\tfrac{1}{2}\beta)$, so that

$$\delta_s = \beta + \tan^{-1}(\tan \tfrac{1}{2}\beta). \tag{17.3}$$

Since β is in practice a small angle, we may put $\delta_s = \frac{3}{2}\beta$, and the angular width of the polar cap is therefore

$$2\beta \approx \frac{4}{3}\left(\frac{2\pi a}{c}\right)^{1/2} P^{-1/2}. \qquad (17.4)$$

For a typical pulsar, with $P = 1$ s and $a = 10$ km, we find the width $2\beta \approx 1°$. This is narrow in comparison with the observed width, which is of the order of $10°$, but several factors may conspire to increase the width. For example, the cone angle containing the emitted radiation may be determined by the angular spread of field lines, $2\delta_s$, rather than by 2β. There may also be some distortions of the field lines from the simple dipole, particularly in the more realistic case in which the field axis is not aligned with the rotation axis. Furthermore, for this inclined field the apparent angular width is increased by a factor cosec α, where α is the inclination angle. We should not therefore be too discouraged by the order of magnitude discrepancy in this calculation.

Some further discouragement does, however, come from the prediction of (17.4) that the angular width should vary with period as $P^{-1/2}$. Although there is some tendency for the largest profile widths to be observed in the shortest period pulsars (Chapter 8), there is practically no support here for the simple polar cap theory. Furthermore, the prediction that the width should vary as $P^{-1/2}$ should apply more or less to any realistic model with an inclined axis, since the defining line of force must in any reasonable geometry be nearer the pole for a pulsar whose velocity-of-light cylinder has a larger radius, i.e. a pulsar with a longer period. It seems unlikely that the scatter of points is concealing a simple relationship of this kind.

Another possible location for the sources is close to the surface of the velocity-of-light cylinder itself. Here the defined range of longitudes may be related to one of several features of the magnetic field configuration, all of which are expected to have an angular width which is not a function of angular velocity, in accordance with observation. In the absence of a theory giving the field configuration close to the velocity-of-light cylinder, we can only indicate which regions might be involved from a sketch of the vacuum field configuration.

The equatorial section of a rotating orthogonal dipole field is shown in Fig. 17.2. Relativity distorts the field into a trailing pattern, so that a straight field line through the magnetic poles crosses the velocity-of-light cylinder at a trailing longitude of 1 rad, where it is inclined at $45°$ to the radius vector. The inclination ϕ' of field lines at other longitudes l,

17.1 Width of the integrated profile

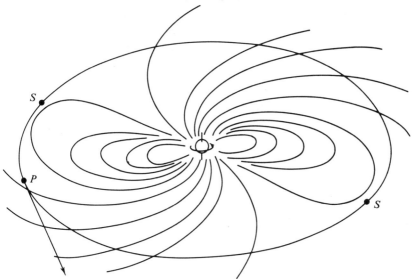

Fig. 17.2. Equatorial field pattern of a rotating dipole, with rotation axis perpendicular to the dipole moment. The point P on the velocity-of-light cylinder is on a tangent to the observer. Points S are typical sources of radiation, which are seen by the observer as they pass through or close to P (see Section 17.4).

measured around the velocity-of-light cylinder from this point is given by

$$\tan \phi' = 1 + \tan \phi \qquad (17.5)$$

where ϕ is the inclination of the field line in a non-rotating field, i.e.

$$\tan \phi = \tfrac{1}{2} \tan l. \qquad (17.6)$$

A possible location for the emitter might, for example, be along a tangential field line, where it remains within a certain small distance of the velocity-of-light cylinder. A more plausible suggestion locates it on the open field lines that cross the velocity-of-light cylinder pointing forwards, in the direction of rotation. A particle flowing outwards and constrained to follow such a line of force must exceed the velocity of light if it is to cross the cylinder; in reality it will acquire sufficient energy to break away from the field line. The region in which this occurs is about 10° long, matching the observed width of the pulse profile. It seems reasonable to suggest that an actual magnetosphere might contain similarly well defined regions in which high-energy electrons are to be expected; further, the width of the region should not depend markedly on period or even on the inclination of the magnetic axis, so that the integrated pulse width will cover a small and constant angular range, as observed.

The emission mechanism: II

17.2 The shape of the integrated profile

The integrated profiles have a marked tendency towards symmetry, which is most noticeable in the more complex profiles. This suggests that the location of the emitting region has some degree of symmetry, favouring in particular the polar cap model (Fig. 17.1). However, there is no sufficiently complete theory of the magnetosphere to rule out any other location on grounds of symmetry alone.

The tendency of the integrated profiles to widen at lower frequencies, which is evidenced particularly by the double-humped profile, favours a location on a curved line of force tangential to the velocity-of-light cylinder, when the lower frequency radiation might be expected to originate at points on the field line further from the tangential point. This possibility will be referred to again when we consider the spectrum of the radiation (Chapter 18).

We conclude, so far, that the characteristics of the integrated profile do not give a clear indication of the location of the source.

17.3 Sub-pulse width

The observations suggest that the sub-pulses represent a beam of radiation from a single source, sweeping across the observer as the neutron star rotates. The beam is typically about 1° wide, symmetrical, and approximately Gaussian in shape. The very high degree of polarisation, and the persistence of the source in the drifting phenomenon, suggest that the source is compact and well organised. The main problem in interpretation comes from the observed independence of sub-pulse width from wavelength.

The generation of a narrow beam of highly polarised radiation is a familiar problem in radio engineering, but the problem of achieving a beam of constant angular width over a range of wavelengths covering several octaves, using a single antenna, has not been solved. The simple reason is that diffraction from an aperture, or from an array of radiators, produces a beam whose width is inversely proportional to wavelength. The observed independence of pulse width on wavelength requires that the beam should be formed in some other way. Some theories for this are discussed in the next chapter; here we expound a purely geometrical solution known as the relativistic beaming effect.

17.4 Relativistic beaming

In this theory, proposed by Smith (1970, 1971) and discussed also by Zheleznyakov (1971), McCrae (1972) and Ferguson (1971), the beam is formed by the relativistic motion of a source located near the velocity-of-

17.4 Relativistic beaming

light cylinder. The source, co-rotating with the pulsar magnetosphere, moves with a velocity approaching the velocity of light; typically a velocity factor $\beta = v/c$ of about 0.8 or 0.9 is necessary in the theory. The relativistic velocity is not in itself the cause of the radiation, which might be emitted by any process which would still operate within the rotating frame of reference.

The high velocity of the source has a remarkable effect on the radiation. In simple terms the radiation is concentrated in a beam along the direction of motion, the beamwidth being approximately equal to Γ^{-1}, where the relativistic factor Γ is given by

$$\Gamma = (1-\beta^2)^{-1/2}. \tag{17.7}$$

For the pulsar, there is a further effect due to a time compression by a factor Γ^2 when the source is travelling towards the observer, so that the beam sweeps across the observer in a time τ given approximately by

$$2\pi\frac{\tau}{P} \approx \frac{1}{2\Gamma^3}. \tag{17.8}$$

The observed widths of typical individual pulses may easily be explained with quite modest values of Γ, between 2 and 3. The beaming effect is calculated in more detail below.

The source itself need not, of course, have an isotropic radiation pattern, so that the observed pulse shape may be the product of a relativistic beam and a polar diagram fixed in the magnetospheric frame of reference. We shall later consider the polarisation characteristics of the sub-pulses in the same way, using the relativistic theory to relate the observed changes of polarisation to the actual polarisation in the radiation polar diagram.

The theory of relativistic beaming may be set out as follows. Suppose an isotropic source is in an orbit with radius r such that the velocity $v = \omega r$ is close to c. An observer, stationary with respect to the centre of the orbit and at a large distance from it, measures the intensity of the source as a function of time. What does he see?

We need to establish relations between orbital position and observed time, and between intensities in the two frames of reference.

Radiation emitted at position $\theta = \omega t$, measured from the radius perpendicular to the line of sight, travels a distance to the observer which is smaller than the distance from the centre of the orbit by OA in Fig. 17.3. Taking a time reference t as measured by a clock at the centre of the orbit,

The emission mechanism: II

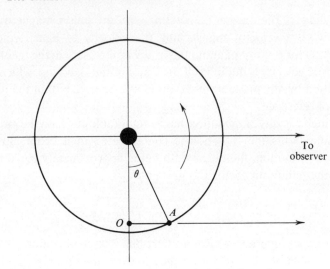

Fig. 17.3. Co-ordinate system for relativistic beaming.

the arrival time T at the observer is

$$T = t - \frac{r}{c} \sin \omega t \sin i \qquad (17.9)$$

where i is the angle between the pole of the orbit and the line of sight. The observer measures time in terms of a phase or longitude $l = \omega T$, so that

$$l = \theta - \beta \sin \theta \sin i. \qquad (17.10)$$

(When θ is small, and i is large, so that the source is moving approximately towards the observer, an approximation may be made:

$$l \approx \theta(1-\beta) \approx \tfrac{1}{2}\Gamma^{-2}\theta. \qquad (17.11)$$

This approximation shows the relativistic time compression already referred to.)

The source is fixed in a frame f' which is instantaneously moving with velocity v relative to frame f, which contains the observer. Time t' measured by a clock in f' is related to t by

$$t = \Gamma t'. \qquad (17.12)$$

Let a photon be emitted at time t', at an angle α' to the source velocity vector. This angle is measured as α in frame f; by the usual aberration formula

$$\cos \alpha' = \frac{\cos \alpha - \beta}{1 - \beta \cos \alpha}. \qquad (17.13)$$

17.4 Relativistic beaming

Consider a number of photons emitted at time t', within an element of solid angle $d\Omega'$. These are seen in frame f to be in solid angle $d\Omega$ where

$$\frac{d\Omega'}{d\Omega} = \frac{\sin \alpha' \, d\alpha'}{\sin \alpha \, d\alpha}. \tag{17.14}$$

Combining (17.13) and (17.14), and using the relation

$$\cos \alpha = \cos \theta \sin i \tag{17.15}$$

we have

$$\frac{d\Omega'}{d\Omega} = \Gamma^{-2}(1 - \beta \cos \theta \sin i)^{-2}. \tag{17.16}$$

The fluxes N, N' of photons in the two frames of reference are related by

$$N \, d\Omega \, dt = N' \, d\Omega' \, dt' \tag{17.17}$$

so that their ratio is

$$\frac{N}{N'} = \Gamma^{-3}(1 - \beta \cos \theta \sin i)^{-3}. \tag{17.18}$$

We now obtain the ratio of the flux densities S, S' in the two frames from

$$\frac{S \, d\nu}{S' \, d\nu} = \frac{N\nu}{N'\nu'} \tag{17.19}$$

giving

$$\frac{S}{S'} = \Gamma^{-3}(1 - \beta \cos \theta \sin i)^{-3}. \tag{17.20}$$

Similarly, the total flux of energy F, F' in the two frames is related by

$$\frac{F}{F'} = \Gamma^{-4}(1 - \beta \cos \theta \sin i)^{-4}. \tag{17.21}$$

When the radiation follows a power law spectrum, so that $S'(\nu') = E\nu'^{\varepsilon}$, then

$$S(\nu) = E\Gamma^{\varepsilon-3}(1 - \beta \cos \theta \sin i)^{\varepsilon-3}\nu^{\varepsilon}. \tag{17.22}$$

The observed radiation has the same spectral index as the radiated spectrum.

The enhancement of flux density over the isotropic value, i.e. the value for $\beta = 0$, is

$$\frac{S}{S_0} = \Gamma^{\varepsilon-3}(1 - \beta \cos \theta \sin i)^{\varepsilon-3}. \tag{17.23}$$

Values of this ratio are plotted in Fig. 17.4 for a source with $\beta = 0.8$, viewed from the plane of the orbit, i.e. for $i = 90°$, and with typical values of spectral index ε (from -1 to -3).

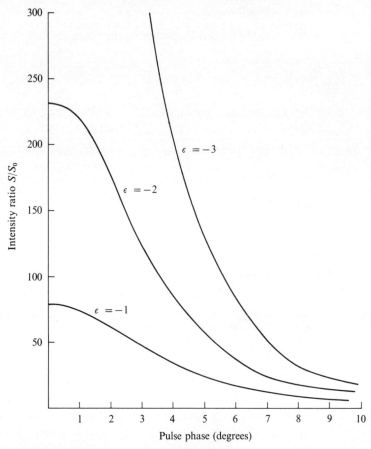

Fig. 17.4. Pulse shapes in relativistic beaming.

The effectiveness of relativistic beaming is shown by the large values of the ratio S/S_0 in these curves. In many pulsars it is known that the level of radiation well outside the phase range of the integrated profile is very small. It is interesting therefore to compare S at $l = 0°$ and at, say, $l = 30°$. For $\beta = 0.8$ and $\varepsilon = 0.2$ this ratio is $273:1$. The 'back-to-front ratio', for $\theta = 0°$ and $180°$, is close to $10^4:1$. This ratio increases rapidly with increasing Γ.

The sub-pulse width may be specified as a full width to half power, $w_{1/2}$. Assuming the sub-pulse is formed entirely by relativistic beaming, with

17.5 Sub-pulse polarisation

an isotropic intrinsic radiation pattern, the width is given approximately by

$$w_{1/2} \approx a\Gamma^{-3} \tag{17.24}$$

where a depends on the spectral index ε:

$$a^2 = 2^{(1/2-\varepsilon)} - 1. \tag{17.25}$$

The observed values mainly lie in the range 1° to 3°, which for $\varepsilon = -2$ corresponds to $\Gamma = 2.9$ to 2.1, or $\beta = 0.94$ to 0.88. The beaming effect is relatively insensitive to the inclination i. This fact may also be expressed in terms of a beamwidth in the polar direction rather than equatorial. This beamwidth is approximately $2a\Gamma^{-1}$, i.e. inversely proportional to the first power of Γ rather than the cube. For a typical value $\Gamma = 2.5$ the beamwidth is 21°, which means that about one-third of the sky is swept by the narrow pulsed beam. Consequently about one-third of such pulsars, with rotation axes oriented randomly with respect to the observer's line of sight, would be observed with at least half the maximum intensity. This result is, of course, very important in determining the population of pulsars in the Galaxy. Relativistic beaming is the only beaming process which produces a fan beam of radiation aligned in this way; the confirmation of this explanation of the pulse formation would therefore involve a reduction of the calculated population by almost an order of magnitude as compared with any other theory.

17.5 Sub-pulse polarisation

It will be recalled that within a sub-pulse the polarisation is typically very high, and that it changes smoothly through a range of elliptical forms in which the hand may reverse and the major axis may rotate by several radians. If the sub-pulse is in fact formed by relativistic compression, then the polarisation at any part of the pulse represents the polarisation at a definite angle of emission in the moving frame of the source; there is no change in the degree of polarisation or the ellipticity in the relativistic transformation. The observed polarisation may therefore be interpreted as the polarisation along a cross-section of the emission polar diagram, covering about 2 radians.

The behaviour of polarisation characteristics in a relativistic transformation has been considered by Cocke & Holm (1972). They show that for a suitable set of axes the relative Stokes parameters V/I, Q/I, U/I are invariant. The set of axes are such that the ray is along one axis in each frame, and another orthogonal axis is common to both frames and perpendicular to the ray in both frames. For the case under consideration,

The emission mechanism: II

where the source is in orbit so that its relative velocity is continually changing, the common axis also changes continuously. However, since there is at all times a set of axes for which the relative Stokes parameters are invariant, the degree of polarisation and the ellipticity are not changed in the transformation. The only change in polarisation is the position angle of the ellipse, which may be evaluated after an awkward geometrical transformation (Ferguson, 1973).

The remarkably simple patterns of polarisation often found in individual sub-pulses therefore correspond to equally simple patterns in the emission polar diagrams, even though these now cover a large angular width. A simple case of this kind is found in PSR 0329+54, where the emission polar diagram appears to be no more complex than that of cyclotron radiation (Fig. 15.1). The Poincaré sphere representation of the observed pulse polarisation is presented in Fig. 9.7. These observed tracks correspond well with tracks obtained in traversing 180° of the cyclotron polar diagram.

Many studies of polarisation have concentrated on the linear component, and particularly on the rate of swing of the position angle. The geometry of this will be examined in the next section, without including the effects of relativistic beaming. It is sufficient to point out here that the rate of swing of position angle is increased by a factor close to Γ^3 in relativistic beaming, and that in addition the geometry of the 'common axis' must be included in any full analysis.

17.6 Position angle of polarisation

There are two ways of explaining the observed changes of polarisation position angle through a pulse. First, it may be that different emitting regions are seen in time succession, these different regions having a range of intrinsic position angles. This explanation generally applies to the integrated profiles. Second, it may be that an individual source, with a fixed intrinsic polarisation, is seen from successively different aspects, so that its apparent position angle changes. This model, known as the single vector model, must be applied to the sub-pulses. We have already discussed the effect of rotation in the single vector model in connection with relativistic beaming.

The two origins of rotation may become confused in actual integrated profiles, since the sub-pulse width may be almost as wide as the integrated profile. This is seen especially in PSR 0833−45 and PSR 1642−03, where the successive pulses are almost identical and superposition produces a highly polarised integrated profile very like the individual pulses. Where the sub-pulses are comparatively narrow, then the integrated

17.7 Single vector model

profile has a lower polarisation. The position angle at any point is then the typical position angle of a sub-pulse at its maximum intensity at that point. The rotation of position angle between successive points on the integrated profile is then unrelated to the rotation in a single sub-pulse; in fact, for PSR 0809+74 the rotation in the sub-pulses is in the opposite sense from the rotation in the integrated profile (Smith, 1974), demonstrating clearly that the two origins of rotation should be considered separately.

17.7 Single vector model

We first examine the way in which the apparent polarisation of a single, linearly polarised source changes as it rotates relative to an observer. This is the 'single vector' model which has been applied particularly to the optical polarisation of the Crab Pulsar. The same model might be applied to the radio sub-pulses of other pulsars. In this model the polarisation position angle is determined by a simple geometry. In Fig. 17.5 the vector

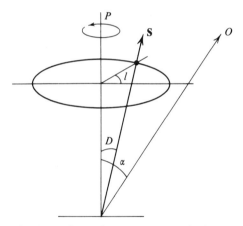

Fig. 17.5. Co-ordinate system for single vector model of polarisation.

S represents an axis within the source, as for example a magnetic field line through the source. The source rotates round the polar axis P, and the observer is at O. The apparent position angle χ of the source vector, measured with respect to the axis of rotation, is given by

$$\tan \chi = \frac{\sin D \sin l}{\sin \alpha \cos D - \cos \alpha \sin D \cos l}. \qquad (17.26)$$

If the source is located near the surface of the pulsar, so that there are no relativistic effects, then the longitude l increases linearly with time.

The emission mechanism: II

The simplest test of this model is provided by the maximum rate at which the position angle changes, which occurs at longitude $l = 0$:

$$\left|\frac{d\chi}{dl}\right|_{l=0} = \frac{\sin D}{\sin(\alpha - D)}. \tag{17.27}$$

As expected, the rate becomes very high when α approaches D, i.e. when the vector **S** sweeps almost across the observer's line of sight; in such a case the position angle may change almost instantaneously through 180°. For example, Wampler, Scargle & Miller (1969) applied the model to the optical polarisation of the Crab Pulsar, and suggest that for this pulsar the angle $(\alpha - D)$ is only a few degrees (we shall later argue that the model is not directly applicable to this case). Again, Radhakrishnan et al. (1969) applied the model to the radio emission from PSR 0833−45, in which the high polarisation of the integrated profile suggests that the model may be applied directly to the integrated profile without reference to individual pulses. The rapid swing of position angle for this pulsar again indicates a near alignment between the magnetic axis and the line of sight at the centre of the pulse.

These interpretations are, of course, not applicable if the time scale of the pulses is compressed by relativistic beaming. In particular, the rapid swing of position angle does not imply a near alignment between the magnetic axis and the line of sight.

17.8 Relation to a rotating magnetic field

If, in contrast to the single vector model, we associate each part of an integrated profile with a definite longitude, then the sweep of polarisation position angle in the profile may be matched to models in which a range of vector directions occurs along a line of longitude. In this case the sweep rate $d\chi/dl$ might represent the changing projection of a series of field lines, or even the curvature of a single field line. Lyne, Smith & Graham (1971) related the observed sweep rates to a model in which the magnetic dipole is perpendicular to the rotation axis, and in which the radiation originates at some fixed radial distance, within the velocity-of-light circle. Radiation is observed when an emitting source is in a special position in relation to the observer, e.g. close to the tangential point. The progress of the emitting sources through this point brings to the observer a changing view of the rotating magnetic field, and the polarisation position angle rotates according to the departure of the field from the radial direction.

17.9 Conclusions: location of the emitter

If the line of sight is at angle α to the rotation axis then:

$$\frac{d\chi}{dl} = -\frac{\cos \alpha}{2\cos^2 l + \frac{1}{2}\sin^2 l \cos^2 \alpha} \qquad (17.28)$$

where l is measured from the magnetic axis.

In this case χ is a monotonic function of l, with the rate $d\chi/dl$ varying between $-(2/\cos \alpha)$ and $-(\frac{1}{2}\cos \alpha)$ but never reversing. The rotation is in the opposite sense to the rotation of the star. The maximum rate occurs when the field lines are tangential, i.e. 90° in longitude away from the dipole axis.

At radial distances approaching the velocity-of-light cylinder, the field configuration is swept back from the simple dipolar field by an angle approaching 1 radian. The value of $d\chi/dl$ then swings through a larger range, but the maximum and minimum values still occur close to the polar and the tangential regions of the field respectively.

The observations summarised in Chapter 16 show: (i) pulsars with simple integrated profiles have simple and smooth variation of χ, indicating a simple field configuration at the emitter; (ii) typical phase rates $|d\chi/dl|$ are in the range 1 to 5. For a single vector model, without relativistic compression, this would mean that the nearest approach of the field axis to the line of sight would lie in the range 50°–10°. For the more probable second model, the observation of phase rates in excess of unity favours locations for the emitter at longitudes where the field lines are nearly tangential rather than normal to the velocity-of-light cylinder.

17.9 Conclusions: the location of the emitter

The two theories for the location of the emitting region are the polar cap theory, in which the source is close to the star surface and restricted in extent by the geometry of the polar cap, and the relativistic beaming theory, in which the source is located close to the velocity-of-light cylinder and restricted in extent by a less obvious feature of the magnetosphere. The appeal of the polar cap theory has been obvious in the large number of papers which favour it; probably the reason is the simple identification of a region of limited angular extent inside which charged particles must be streaming out from the surface. As we see in the next chapter, this streaming leads to a simple suggestion for the radiation mechanism, namely curvature radiation. On the other hand, the observational evidence both of the radio emission and of the whole spectrum of emission from the Crab Pulsar favours the relativistic beaming hypothesis, especially the evidence on the pulse widths, the high degree of polarisation, and the rate of swing of polarisation.

There is as yet no model of the magnetosphere which applies to the orthogonal or inclined magnetic field. It will be interesting to see if such models can indicate regions of special interest, as for example an irregularity in the field lines or a concentration of energy which might be identified with the seat of the radiation. In the aligned rotator the hydrodynamic model of Kuo-Petravic *et al.* (1974) shows that the field lines do not contain irregularities, but that there is a marked concentration of energy in equatorial regions near the velocity of light cylinder; if this is also generally true for inclined fields the relativistic beaming hypothesis would be favoured.

The choice of location must therefore be kept open. Theorists may prefer to develop the polar cap model, because of its simplicity. Observers may prefer to examine the evidence of the actual pulses, which have so far only been fitted in detail to the relativistic beaming model, and attempt to devise further and more critical experiments on which the decision can be made. The author of this book works from the observational side; he claims that this strongly favours a location close to the velocity-of-light cylinder.

References

Cocke, W. J. & Holm, D. A. (1972). *Nature Phys. Sci.* **240**, 160.
Ferguson, D. C. (1971). *Nature Phys. Sci.* **234**, 86.
Ferguson, D. C. (1973). *Astrophys. J.* **183**, 977.
Kuo-Petravic, L. G., Petravic, M. & Roberts, M. (1974). *Phys. Rev. Lett.* **32**, 1019.
Lyne, A. G., Smith, F. G. & Graham, D. A. (1971). *Mon. Not. R. astron. Soc.* **153**, 337.
McCrae, W. H. (1972). *Mon. Not. R. astron. Soc.* **157**, 359.
Radhakrishnan, V., Cooke, D. J., Komesaroff, M. M. & Morris, D. (1969). *Nature, Lond.* **221**, 443.
Smith, F. G. (1970). *Mon. Not. R. astron. Soc.* **149**, 1.
Smith, F. G. (1971). *Mon. Not. R. astron. Soc.* **154**, 5P.
Smith, F. G. (1974). *Mon. Not. R. astron. Soc.* **167**, 43P.
Wampler, E. J., Scargle, J. D. & Miller, J. S. (1969). *Astrophys. J.* **157**, L1.
Zheleznyakov, V. V. (1971). *Astrophys. Space Sci.* **13**, 87.

18
The emission mechanism: discussion

Starting on a discussion of the many proposals that have been made to explain the radio and optical pulses is like entering a quagmire, in which each step takes one deeper into unpleasant complications, or like entering a minefield, in which each step may reveal a new factor compelling retreat. The early days of pulsar theories naturally swept all the difficulties away; everything could be explained by the simple ideas of synchrotron radiation and curvature radiation, with suitable modifications to account for coherence. The complications of plasma waves soon removed the simplicity, and it was realised that there is not only a problem in the emission but also in the propagation of the radio waves through a very unusual magnetoionic medium. Even with a partial understanding of the propagation problems, which amounted essentially to assuming that there were no dispersive effects in the pulsar magnetosphere, the most intractable problem remained: why do we observe pulses at all?

It now might seem from a casual glance through the literature that every conceivable radiation process occurs in pulsars. Perhaps they all do; however, it is best to start not by postulating more processes but by looking closely at the observations and from them describing the conditions within the source. This will at least narrow the range of relevant processes, although it will not of course account for the existence of the particle energies and distribution that it will reveal.

We first review the experimental evidence concerning the location of the emitting region. Then, keeping close to observational data, we find volume emissivities and discuss energy densities. Finally, the emission mechanisms themselves are discussed.

18.1 Location

The location of the emitter is obviously physically determined by the configuration of the magnetic field. The simplest and most obvious location is on field lines passing through the polar caps, which define suitably small regions near the surface. There are, of course, other locations defined by the magnetic field, not necessarily close to the surface. These we examine later. The main evidence presented for

The emission mechanism: discussion

theories in which the emission is from near the surface, where charged particles streaming along the field lines may be expected to emit curvature radiation, is as follows:

(1) The theory accounts for the duty cycle of the pulsation. Since the particles stream along the field lines, the emission would be directed into a cone at each pole; for an oblique rotator this accounts for a single observed pulse, while for a suitable alignment of the rotation axis and the direction of the observer the theory can also account for interpulses.

(2) The linear polarisation, both optically and for radio, swings in position angle in an appropriate way, corresponding to the varying projected angle of the field lines as seen by the observer.

Both these pieces of evidence are not quite straightforward. The details of the pulse profiles and the interpulses do not reveal quite such a symmetrical situation as in the simple model, while the existence of sub-pulses with their own peculiar polarisation behaviour, including strong circular components and rapid swings of position angle, does not support the 'single vector' model of polarisation (Chapter 17).

A location near the velocity-of-light cylinder involves pulse formation by the relativistic beaming process (Chapter 17). This is a purely geometrical process which gives a pulse beamwidth independent of frequency. The main evidence for relativistic beaming is:

(1) The width of sub-pulses, which seem to represent the basic beam of radiation, does not depend on radio frequency.

(2) The integrated profile of the Crab Pulsar is very similar over 40 octaves.

Neither of these observational facts can be accounted for by any theory involving a beamed radiation process; they essentially require the purely geometrical approach of the relativistic beaming theory. As we will note later, relativistic beaming can account for drifting sub-pulses, although these are not to be regarded as such direct evidence as the above two points.

An entirely different argument, due to Goldreich, Pacini & Rees (1971) also places the optical radiation from the Crab Pulsar far from the surface. The minimum brightness temperature for an optical source at the surface, where the projected area of the source must be less than the projected area of the pulsar surface, is 10^{15} K. There is no known possibility of obtaining coherent optical radiation, nor is there any observational evidence for it in, for example, a variability of the optical pulse power. The brightness temperature can therefore be related to a

18.1 Location

minimum energy for the radiating particles. Presumably these are electrons or positrons, for which the energy must be at least the thermal energy corresponding to 10^{15} K. This means that their energy factor γ is greater than 10^5. Emission by synchrotron or curvature radiation, peaks at a frequency $\nu \approx (\gamma^3 c)/R$ where R is the radius of curvature of the electron orbit. If $\gamma > 10^5$, this implies $R > 10^{10}$ cm. The electrons therefore travel nearly in straight lines, and it is easy to show that they lose only a fraction $\leq 10^{-9}$ of their energy before they leave the pulsar. But the observed power in the optical radiation is already about 10^{-4} of the available energy supply from the rate of loss of rotational energy, so that there is a deficit by a factor of at least 10^5 in the energy required for radiation at such a high brightness temperature. The optical pulses cannot, therefore, be generated in a small region near the surface; instead they must be generated in a larger region, further out.

Shklovsky (1970) presents a similar argument, showing that the optical and X-ray emission can only come from electrons close to the velocity-of-light cylinder. The electron energy factor γ then need only be 10^2, with a magnetic field of about 10^4 gauss. The only escape from these arguments is to postulate that the radiation is generated by protons rather than electrons. This requires a theory in which a large proportion of available energy goes into the protons rather than the electrons, which seems unlikely.

The similarity of the integrated profiles of radio, optical and X-ray radiation is strong evidence that the source is similarly located over the whole spectrum. The argument that the optical radiation originates close to the velocity-of-light cylinder therefore supports the previous argument based on the radio sub-pulse structure.

The location close to the velocity-of-light cylinder must be related to some feature of the magnetic field. Lyne, Smith & Graham (1971) point out that there is a suitable range of longitudes defined by the region in which open field lines cross the velocity-of-light cylinder pointing in a forward direction, so that particles cannot stream outwards along the lines without travelling faster than the velocity of light. This might therefore be a region in which there is an accumulation of high-energy particles. Evidence for this location is provided by the polarisation in the integrated profile of several pulsars. Fig. 17.2 indicates the region, on a sketch of an orthogonal dipole field in which it is assumed that the flow of charged particles has no effect on the field. Lyne *et al.* (1971) were able to relate the rate of change of position angle of the field lines across this region with the rate of swing of polarisation position angle through the integrated profile.

The emission mechanism: discussion

18.2 Energy density

The volume emissivity for the observed pulses is astonishingly high. Whatever the radiation mechanism may be, and wherever the source is located within the magnetosphere, the observed radiation must come from a volume which is smaller than a sphere with radius c/Ω, i.e. the radius of the velocity-of-light cylinder. The very rapid signal fluctuations observed from PSR 0950+08 imply much smaller dimensions. Typical values for the volume emissivity for radio pulses (Smith, 1973) are 10^8 W m^{-3} for the Crab Pulsar and 10^4 W m^{-3} for PSR 0329+54. The optical and X-ray powers for the Crab Pulsar are larger, reaching at least 3×10^{12} W m^{-3}, i.e. 3×10^{13} erg s^{-1} cm^{-3}.

These high power densities reach beyond the range of terrestrial experience, where thermal powers of the order of megawatts per cubic metre are becoming familiar in nuclear reactors; on a smaller physical scale, a similar power density is known in radio oscillations in high-power klystrons and, again much smaller, in some solid-state oscillators. The corresponding radiation energy density for the optical and X-ray radiation from the Crab Pulsar is 10^3 erg cm^{-3}; this is small compared with the energy of the magnetic field $B^2/8\pi$ when $B \approx 10^6$ gauss, as expected near the velocity-of-light cylinder. Although the energy of the emitting particles is large, it is still not large enough to disturb the configuration of the magnetic field. (There is, however, a suggestion by Manchester, Tademaru, Taylor & Huguenin (1973) that the energy of some of the largest radio pulses from PSR 0950+08 violates this condition; the observational evidence for this is not yet strong enough to sustain an argument that the source must therefore be placed closer to the surface where the field is stronger.)

The field strength of the radio emission is also higher than in any other astrophysical situation. For the Crab Pulsar it reaches 10^9 V m^{-1}, in radio waves with a wavelength of 1 m. This immediately implies that no charged particle in the vicinity can have an energy less than about 10^9 eV on average. There can be no low-energy electrons in the magnetosphere in the presence of this radiation, and there is therefore no significance to any values of plasma frequency or gyrofrequency calculated for electrons with non-relativistic energy. It is the existence of these very high field strengths that complicates the theory of propagation in the magnetosphere (Kegel, 1973).

It is clear that the radiating particles must be supplied with energy at a great rate, and that this energy must be well organised. Sturrock (1971) suggests that this energy supply must flow out through the polar caps, since relativistic particles should flow along open magnetic field lines.

This is not, however, the only or even the main energy flow: the hydrodynamic model of Kuo-Petravic, Petravic & Roberts (1974) shows that energy flows out also from equatorial regions; for an aligned rotator their model shows a maximum energy density near the equator at the velocity-of-light cylinder. It seems likely that an oblique rotator will have similar energy concentrations at two opposite points close to the velocity-of-light cylinder, and away from the magnetic poles. These concentrations would presumably correspond to the sources of optical emission in the Crab Pulsar. The radio emission from other pulsars need not, of course, trace out precisely the regions of greatest energy density since it depends more on conditions of coherence in particle motion than on total energy.

18.3 Spectrum

The evidence shows that the spectrum of the Crab Pulsar is continuous from infrared, through optical, ultraviolet and X-rays to gamma-rays. Most of this spectrum is consistent with a synchrotron radiation spectrum from electrons with a suitable range of energies (Zheleznyakov & Shaposhnikov, 1972). The turn-over in the infrared is self-absorption. At higher frequencies the spectral index $\alpha \approx 1.2$, corresponding to a spectrum of electron energies

$$N(E) \propto E^{-\gamma}$$

where

$$\gamma = 2\alpha + 1 = 3.4.$$

The actual values of electron density depend on the magnetic field. In the extreme case of equipartition of energy between the electrons and the field there would be 6×10^{11} electrons cm^{-3} distributed with energy as

$$N(E) = 1.4 \times 10^3 (E_{\text{eV}})^{-3.4}$$

for $E > 3 \times 10^8$ eV. The magnetic field strength would then be 6×10^4 gauss. For a higher field strength the particle density and energy would both be smaller.

The only complication which is expected in this synchrotron spectrum is that part of the gamma-radiation may be due to inverse Compton radiation. This results from scattering of the synchrotron optical and X-ray radiation by very high-energy electrons. It seems likely that this is an important effect, since if only the synchrotron process were operating in the gamma-ray region, this would be removing energy from electrons too fast for it to be replenished. The radiation loss would be so fast that

The emission mechanism: discussion

the electrons would not even complete one turn in their gyration round the magnetic field. Inverse Compton effect is discussed for the Crab Pulsar by Zhelesnyakov & Shaposhnikov (1972).

18.4 Radio frequency radiation

In contrast to the optical radiation, it seems likely that the radio radiation is emitted by a narrow-band, resonant process. This assertion contradicts the theories which involve curvature radiation and synchrotron radiation, and it is important first to argue the point. As we have seen, there are direct indications of frequency structure in single pulses from the Crab Pulsar and in some others (Chapter 9); perhaps the strongest argument, however, is the variation of integrated pulse shapes with frequency, which suggests at least some varied and steep spectral features. But at the same time there is obviously a broad-band nature to the radio pulses, so that an assertion of a basic narrow-band mechanism must imply a simultaneous excitation of a wide range of resonant frequencies.

The most telling argument for a resonant mechanism is the negative argument that no proposal has yet emerged for a broad-band mechanism which has the observed polarisation properties, and which has in particular 100% elliptical polarisation over a wide range of frequencies. Synchrotron radiation, for example, can only have 100% polarisation if the radiating charges are perfectly collimated; in this case, however, the radiation would be beamed with a width which is frequency dependent, contrary to observation. The same applies to curvature radiation. It may be, alternatively, that the polarisation is imposed by propagation conditions away from the source; then a theory is required that gives such a similar behaviour over a wide frequency range. But no suitable non-dispersive propagation phenomenon is known.

A resonant source is naturally fully polarised. The similarity of polarisation over a wide frequency range could, however, only be obtained from a corresponding similarity over a range of resonant sources. In theories involving resonance it is therefore necessary to postulate that an ensemble of sources exists, physically connected as they might be if they are strung along a line of force, but with different resonant frequencies covering the whole range of the radio spectrum.

The nature of the resonance itself must be determined primarily by the magnetic field, which is known to dominate all movements of charged particles. The resonant frequency is not simply the Larmor frequency; the charges will have relativistic energies, reducing the gyrofrequency by the factor γ; furthermore, they may be streaming along the field lines, giving a Doppler shift within the magnetosphere. (The observed radiation will

18.4 Radio frequency radiation

also have a Doppler shift due to the velocity of the source as a whole, if relativistic beaming is occurring.)

The simplest hypothesis is that the resonance is a pure gyroresonance, at angular frequency $\omega_H = eH/\gamma mc$. There must also be coherence in the particle motion, so that bunches with a similar energy gyrate together. We use the theory presented in Chapter 15 to calculate first the electron energies and densities which would account for the radiation if it is the gyrofrequency, and then the incoherent synchrotron radiation expected from the same electrons. This power turns out to be compatible with the observed optical and X-ray power. Typical values, given by Smith (1973) for the Crab Pulsar, are as follows:

(i) Observed frequency 200 MHz, Doppler shifted, due to the motion of the source, from about 50 MHz. Assuming a field 10^5 to 10^6 gauss, corresponding to the outer part of the magnetosphere, the energy of an electron giving this resonant frequency must give $\gamma = 5 \times 10^3$ to 5×10^4.

(ii) At 50 MHz a gyrating relativistic electron radiates 5×10^{-20} W in the fundamental, whatever its energy. The volume emissivity is 10^8 m^{-3}, and the typical dimensions of a coherent bunch of electrons must be about 1 m^3. Consequently a density of N m^{-3} gives an emissivity

$$5 \times 10^{-20} N^2 = 10^8 \text{ W m}^{-3}$$

and $N = 5 \times 10^{13}$ m^{-3}. This is a minimum value for the density, since the coherence may not be complete.

These values are well below the maximum values expected from the magnetosphere theories. For pulsars other than the Crab Pulsar, smaller values apply; in particular the magnetic field in the outer part of the pulsar magnetosphere is smaller, since it falls as the inverse cube of the period if the surface field is the same. The particle energy must therefore also fall so as to keep the gyrofrequency within the radio frequency band.

Relativistic electrons gyrating in this way normally emit very much more power in harmonics than in the fundamental. In this case, however, the harmonics are not enhanced by coherence, because the bunches are large compared with the wavelength. If, in particular, the electron distribution round a circular orbit were a simple sinusoid, there would be no coherence at all except at the fundamental. The total power radiated in the harmonics is given by the synchrotron formulae (Chapter 15). The power is radiated mainly at frequencies γ^3 above the radio frequency, i.e. within the optical and X-ray region for the Crab Pulsar. The volume emissivity varies as γ^4, so that it cannot be predicted with accuracy; however, the observed value of 3×10^{12} W is easily accounted for with

The emission mechanism: discussion

$N = 5 \times 10^{13}$ m^{-3} and $\gamma = 3 \times 10^4$. It is therefore entirely possible that the same electrons are responsible both for the radio and optical emission; only a small proportion of the energy is radiated in the radio region, even with full coherence.

18.5 Concluding discussion

Two schools of theory have emerged from the many and complex discussions on the origin of the radio emission. Unfortunately the two schools also diverge on the geometrical question of the location of the emitter, recruiting the same proponents as they do on the nature and mechanism of the emitter. The divergence on mechanism stems only partly from the geometrical question; there is a general problem for either location in finding a mechanism that is both coherent and broadband. We have already reviewed the evidence on the geometrical location: if it is accepted that the observed sub-pulses are indeed elementary beams of radiation, then the combined facts that they are fully polarised and that their width is independent of radio frequency leads firmly to the conclusion that the emitter is close to the velocity-of-light cylinder. Theoretical work ought now to be addressed to the separate question: is the radiation inherently narrow-band or broad-band? And if it is narrow-band, how can there be a sufficiently widespread set of resonant frequencies all simultaneously excited, so that a broad-band pulse is observed?

One remaining geometrical question should, however, be disposed of before the mechanism is discussed. This concerns the interpulses which are observed in only a few pulsars. Any theory which invokes relativistic beaming must accommodate an asymmetry in all pulsar magnetospheres, which removes the inherent symmetry of an inclined dipolar field and prevents or distorts the radiation from one half of the field. How can such an asymmetry arise, especially in the outer parts of the magnetosphere where the dipole field will be relatively undistorted by higher order spherical harmonic components of the field?

The Crab Pulsar provides an interesting commentary on this question. On the one hand, the optical radiation is fairly symmetric between the main pulse and the interpulse (except for an asymmetry in the spacing between the pulses). On the other hand, the radio radiation at long wavelengths, where a precursor appears and where the polarisations of the main pulse and interpulse differ, does not show the same symmetry. The comment is obvious: a fairly symmetrical situation, as revealed by the optical pulses, gives rise to asymmetrical radio pulses.

A possible solution to this puzzle is to suppose that the excitation which gives rise to the coherence in the radio radiation is more sensitively

18.5 Concluding discussion

dependent on small features of the field than is the synchrotron radiation which is observed optically. For example, the coherence might depend on conditions at the polar caps: it might depend on particle streams which can be asymmetric, as indeed would be asserted by the 'polar cap' theorists who wish to explain asymmetries in the observed interpulses, such as the polarisation behaviour of PSR 0950+08.

Evidently one feature of the relativistic beaming theory is its flexibility; since it does not depend on a precise description of the radiation mechanism or of its excitation, there are *ad hoc* explanations of almost any detail of the observed very complex phenomena. Alternatively, and preferably, one should establish the relativistic beaming theory on a logical foundation, and then use it to provide a description of the radio radiation as it would be seen within the rotating magnetosphere. On this basis, the radiation polar diagram is broad, possibly near-isotropic, the polarisation is close to 100% and changes smoothly and very simply over the polar diagram, there are locations for any one frequency spread over a range of longitude, and at any one longitude a range of frequencies can be observed. The range of frequencies does, however, differ at different longitudes, corresponding to the variation of spectrum over the integrated profile; similarly at any one time a single frequency is observed only from a very restricted range of longitudes, corresponding to the limited duration of a sub-pulse. Furthermore, the radiation must be highly coherent.

Coherence may be obtained either by a geometric condition within the source, such as an arrangement of charges in bunches or in a sheet, or it may be obtained in a maser amplifier. We turn first to coherence within the source. If the radiated polar diagram is very wide, and the radiation is coherent, then the elementary radiator is only of the order of one wavelength across. Furthermore, the appearance of circular polarisation shows that the bunch is gyrating, and it follows that the radiation is some form of cyclotron radiation. A rough numerical analysis, as shown in this chapter, provides reasonable values for electron energy, electron density and magnetic field in such a theory.

The explanation of coherence by a maser action is more difficult to sustain. Maser action is inherently a resonance, so that the theory must provide resonant amplification over a wide range of frequencies. But the amplified radiation is also beamed, so that one must now suppose that relativistic beaming is not after all responsible for the pulses, and that the maser beaming effect is practically independent of frequency. Furthermore, the amplification, and the beamwidth, must also be independent of polarisation. No description of such a maser process has been given, and indeed it seems unlikely that such a process can be found.

The emission mechanism: discussion

To sum up, the relativistic beaming of coherent cyclotron radiation provides the only explanation so far found for the radio pulses. The theory may be extended to cover the synchrotron radiation observed optically in the Crab Pulsar, and it accounts well for the existence of pulses over 40 octaves of the electromagnetic spectrum. How the coherent motion comes about is not at all understood; it represents a department of plasma physics which we would very much like to simulate in the laboratory, but which we scarcely know how to describe.

References

Goldreich, P., Pacini, F. & Rees, M. J. (1971). *Comments on Astrophys. Space Sci.* **3**, 185.

Kegel, W. H. (1973). *Astron. Astrophys.* **22**, 475.

Kuo-Petravic, L. G., Petravic, M. & Roberts, K. V. (1974). *Phys. Rev. Lett.* **32**, 1019.

Lyne, A. G., Smith, F. G. & Graham, D. A. (1971). *Mon. Not. R. astron. Soc.* **153**, 337.

Manchester, R. N., Tademaru, E., Taylor, J. H. & Huguenin, G. R. (1973). *Astrophys. J.* **185**, 951.

Shklovsky, I. S. (1970). *Astrophys. J.* **159**, L77.

Smith, F. G. (1973). *Nature, Lond.* **243**, 207.

Sturrock, P. A. (1971). *Astrophys. J.* **164**, 529.

Zheleznykov, V. V. & Shaposhnikov, V. E. (1972). *Astrophys. Space Sci.* **18**, 166.

19

Supernovae: the origin of the pulsars

Despite the difficulties of explaining the subtleties of the radio pulses, and despite the continuing theoretical problems of the magnetosphere, the crystalline surface, the superfluid interior and the possible solid core, there is no doubt about the identification of pulsars with neutron stars. There is no doubt either about the difficulty of forming a neutron star: it cannot collapse smoothly as part of a gentle aging process, but can only be formed as part of a catastrophic event, which we assume to be a supernova explosion. We would expect, therefore, to find pulsars associated with supernova remains, and we would expect to find the distribution and population of pulsars within the Galaxy to be closely related to the frequency of occurrence of supernovae. These two expectations are the subjects of the next two chapters. We start with a review of the observational data concerning supernovae.

19.1 Supernova explosions

The first telescopic observation of a supernova, made by Hartwig in 1885, was of a star in the Andromeda Nebula, M31. This single star for several weeks became so bright that it emitted more than 10% of the light emitted by the whole of the nebula. All modern observations of supernovae have followed this example in that they refer only to extragalactic events. Most observations have been made in organised photographic patrols of bright extragalactic nebulae, notably by Zwicky (1962). The only accounts of supernova explosions observed in our own Galaxy are four centuries old or more; even so, some accounts are sufficiently detailed to allow close comparisons to be made with the extragalactic observations, particularly in respect of magnitudes and light curves. A detailed account of these and of many other aspects of supernovae has been given by Shklovsky (1968).

At least two types of supernovae can be clearly distinguished. Of these Type I is the most homogeneous, showing remarkable uniformity in luminosity at maximum brightness, in the light curve, and in the spectrum. A typical light curve is shown in Fig. 19.1. After a sharp rise in luminosity there is a flat top lasting a few days, then a drop of 2 or 3

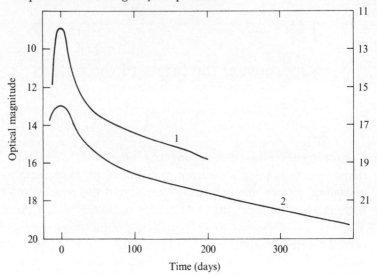

Fig. 19.1. Photographic light curves of Type I supernovae. 1, NGC 4621 (right-hand scale); 2, NGC 1003 (left-hand scale). (After Minkowski, 1964.)

magnitudes in 20 to 30 days, followed by an exponential decay of luminosity with time constant close to 50 days. The spectrum shows only broad spectral features. Type II light curves show generally a broader maximum, a smaller drop after the maximum, then a steeper decline. The spectrum at peak brightness is a continuum, but after the peak it develops emission lines and becomes similar to a nova spectrum.

The most important distinction between Type I and Type II supernova concerns their association with the two different types of stellar population, Population I (the younger disc stars) and Population II (the older halo stars). Unfortunately but perhaps inevitably, Type I is associated with Population II, and Type II with Population I. This relation is best demonstrated in a statistical analysis by Katgert & Oort (1967), who showed (i) that Type II *only* came from spiral galaxies, and particularly from late-type spirals which are predominantly Population I, and (ii) that elliptical galaxies, which contain only Population II, produce only Type I supernovae. This clear distinction shows that the two types come from stars in different ranges of mass. The older Population II cannot contain very massive stars, since these evolve too rapidly and they should already have exploded. The mass of a star which becomes Type I supernova is found from this argument to be less than $1.2 M_\odot$.

Not only are Type II supernovae associated with young stars, which can be massive, but they are also observed to expel large masses in the

19.2 Frequency of occurrence of supernovae

explosion. Expanding gas clouds can be observed in the line spectra of Type II extragalactic supernovae, and more detailed observations are possible for the visible remains of the galactic Type II supernova Cass A. A mass of at least $1M_\odot$ is expelled from a Type II supernova, so that the typical source star must be at least several times more massive than Type I.

The source of energy for a supernova explosion is the gravitational energy released when a star collapses. The collapse can only occur when energy can be radiated rapidly from the core in the form of neutrinos, so that internal pressure is removed and a large part of the star collapses under free fall. The subsequent release of energy blows the outer part of the star apart, forming the supernova shell. The nuclear processes involved were described by Fowler & Hoyle (1964).

It was pointed out by Hoyle, Fowler, Burbidge & Burbidge (1964) that gravitational collapse can easily lead to the gravitational singularity now known as a 'black hole'. This occurs when the collapsing mass exceeds about $2M_\odot$, provided that the collapse is not inhibited by one of several possible braking processes. These are due to rotation, or turbulence, or a magnetic field. Shklovsky (1971) has particularly emphasised the importance of rotation, which can prevent all material beyond a quite small radius collapsing. It follows that even though the mass of a neutron star cannot exceed $1.7M_\odot$ (Chapter 5) such a star could result from the collapsing core of any large supernova, which might have a mass well in excess of $2M_\odot$.

19.2 Frequency of occurrence of supernovae

The occurrence of supernovae is such a rare event in any individual galaxy that a measurement of the average frequency of occurrence requires continued observation of a large number of galaxies. Inevitably the observations are far from continuous, and some supernovae will be missed. The extent of this incompleteness is demonstrated by the statistical work of Katgert & Oort (1967), who revised the previous estimate for the average frequency from one per 450 years (Minkowski, 1964) to one per 40 years.

The evidence from our own Galaxy is less significant, since the assessment of completeness requires some guesswork about the thoroughness of ancient observers. No supernovae have been observed since 1572 (Tycho) and 1604 (Kepler), but others may have occurred in obscured regions of the Galaxy. The prime example is the radio source Cass A, which must have exploded about the year 1700. There is good evidence that this was a Type II supernova. It was not reported from any

Supernovae: the origin of the pulsars

observatory, even though Cass A is circumpolar in the Northern Hemisphere. The visible remnant, discovered after the source of the radio emission had been located with sufficient accuracy, is heavily obscured, to the extent of 10 visual magnitudes. The supernova explosion would then hardly bring the star into the range of naked eye visibility.

The radio remains of other supernova explosions can be distinguished from other discrete galactic radio sources by their size, shape, and spectrum. Over 100 supernova remains (SNR) have been catalogued by Ilovaisky & Lequeux (1972). Woltger (1972) has estimated the lifetime of the remnants, and after allowing for the incompleteness of the catalogues he estimates that they were formed at a rate between one per 60 years and one per 45 years. The typical age is about 5×10^4 years. These supernova remnants are distributed uniformly over a galactic disc with radius $R = 8$ kpc, with a surface density falling by a factor 10 at $R = 12$ kpc. Within $R = 6$ kpc the z-distribution is close to an exponential with scale height 90 pc. The age and distribution of these remnants agree with an origin in Type II supernovae, which in turn are associated with young Population I stars. We shall compare the spatial distribution of pulsars with the distribution of supernovae in Chapter 20.

The most remarkable feature of the observed rate of occurrence of supernovae is that the rates for Type I and II supernovae are very similar, despite the fact that they appear to be distinct categories originating in different types of star. Furthermore, the total energy release for the Type II is much greater than for Type I, as demonstrated by the kinetic energy of the expanding remnants; the optical emission is not, however, very different, so that the chances of observation are similar. The ratio of visible energy to kinetic energy may be 10^3 larger for Type I than for Type II.

19.3 Supernovae in binary systems

It is very remarkable that the neutron stars associated with X-ray sources are all members of binary systems, while more than 100 pulsars were discovered before one was found which was a member of a binary system (PSR 1913+16, see Chapter 7). This is not the result of observational selection; it is quite clear that most pulsars are single and not members of binary systems. Binary systems are, however, common among all types of star, and particularly among the early type stars which are expected to become Type II supernovae. Among these binary systems about 70% have periods shorter than 10 days (Batten, 1967), which would make them particularly easy to detect. The absence of 'binary' pulsars therefore requires an explanation, either in terms of an origin only in solitary stars

19.3 Supernovae in binary systems

or in terms of the disruption of binary systems at the time of the supernova explosion.

A supernova explosion is more likely to disrupt a binary system if it is the heavier star which explodes. Boersma (1961) calculated the effect on the orbits if the heavy star disrupts and throws a proportion of its mass outside the system in a time short compared with the orbital period. The orbits of close binaries are generally circular, so that the calculation depends only on the proportion q of mass left after the explosion. If $q < 0.5$ the binary system is no longer bound, and the two stars will fly apart at typical orbital velocities.

Boersma's analysis is not seriously modified if the material is ejected more slowly, so that the orbit is progressively modified. A more important effect in close binary systems is the collision of the expanding shell with the companion star. Colgate (1970) pointed out that in addition to the direct transfer of momentum in the collision there might also be sufficient heating on the inner side of the companion for it to expel material, tending to disrupt the orbit further by a rocket effect. A comprehensive analysis of the conditions under which the binary would be completely separated was given by McCluskey & Kondo (1971). The limiting conditions, under which there is no disruption, depend on the kinetic energy of the expanding material, but even for the closest binaries there will be no disruption if the original mass of the exploding star is less than 0.2 of that of its companion, or if less than 0.2 of its mass is expelled. Most Type I supernovae are expected to remain bound, since they have a comparatively small mass; furthermore they expel a small proportion of mass (and at a lower velocity) than do Type II.

Type II supernovae should often disrupt a binary system at the time of explosion, leaving an expanding shell of gas and two high-velocity stars, one probably a neutron star. Blaauw (1961) suggested that some stars observed to have high velocities, such as U. Geminorum, could have originated in this way. We shall return to this question when we discuss the high velocity of the pulsars.

Disruption is by no means certain for close binary systems in which a Type II explosion occurs, even though the exploding star may start as a massive star, typically with 20 solar masses, and end as a comparatively light neutron star. The expected disruption is prevented by mass transfer within the binary system. The sequence of events has been described by van den Heuvel & De Loore (1973) for a typical close binary system consisting of stars with mass $25M_\odot$ and $8M_\odot$, with an orbital period of 4 days. Fig. 19.2 shows the sequence. The essential feature is that the heavier star evolves more rapidly, reaching the giant stage before the

Supernovae: the origin of the pulsars

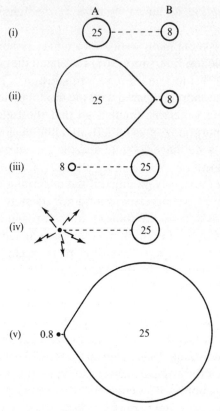

Fig. 19.2. The evolution of a binary X-ray source. (i) In the binary system the more massive star A evolves first. (ii) The more massive star has now expanded, and transfers mass to the lighter star. (iii) Star A now collapses to a small helium star, with 8 solar masses. (iv) Star A now explodes, leaving a neutron star with 0.8 solar masses. The neutron star may be observed as a pulsar. (v) Star B now evolves, expands, and transfers mass to the neutron star, which now becomes an X-ray source. (After van den Heuvel & de Loore, 1973.)

lighter star has appreciably changed. The star can then fill the 'Roche lobe' of the binary system, which is the volume inside which material can orbit round the single star. At this stage mass is transferred rapidly to the secondary star, and their masses practically interchange. The original primary loses all its outer hydrogen-burning layers, leaving only a helium core. This core then evolves rapidly, and explodes; but at this stage the exploding star is the less massive star and disruption of the binary is avoided.

19.4 Chances of observing a binary pulsar

The short-lived helium star stage is recognised as a Wolf–Rayet binary system. After the explosion the primary is left as a neutron star (or possibly as a white dwarf or black hole), orbiting the enlarged secondary star, which now evolves more rapidly, eventually in its turn filling its own Roche lobe and transferring mass back on to the primary. Since the primary is a collapsed object, there is a large release of energy as the mass falls on it. This system then becomes an X-ray source, as described in Chapter 4.

The next stage of evolution does not much increase the mass of the neutron star, which can probably only accrete matter at the low rate of $10^{-7} M_\odot$ yr^{-1} (van den Heuvel & De Loore, 1973). Instead the envelope of the secondary star is blown out of the orbiting system by radiation pressure, and when the secondary is left as a helium star of $8M_\odot$ it is still the heavier star. When it explodes, the binary system will disrupt, and two neutron stars (or other condensed objects) will fly apart.

The outcome of this sequence of events is that a binary pulsar might be expected during a period of 4×10^6 years after the first explosion. Apart from this interlude, binary systems are generally disrupted by supernova explosions. We now examine the statistical chances of observing a neutron star as a pulsar during this binary interlude.

19.4 The chances of observing a binary pulsar

The statistics have been presented by van den Heuvel (1973). A star in a close binary system must have a mass of at least $16M_\odot$ or $18M_\odot$ if it is to evolve into a neutron star, since it must follow the mass transfer process outlined above. If it is essentially solitary, or in a long-period binary system, it must have a mass in the range $4M_\odot$ to $10M_\odot$ (Cameron & Canuto, 1973). In both cases it may evolve instead into a black hole, the choice depending on the presence of a braking mechanism during the collapse.

The ratio of explosion in binary and in solitary systems is therefore about 1%, which agrees very well with the discovery of one binary system in the first hundred or so known pulsars. We should, of course, remark that the statistical significance of one example is not very high; on the other hand, the calculation leading to the ratio of 1% was at least a prediction made before the discovery.

As we have seen, the only means of avoiding disruption is mass transfer, which applies only to close binary systems. Long-period binary systems are therefore not to be expected; again this agrees with observation as far as present tests allow.

Supernovae: the origin of the pulsars

19.5 Velocities acquired by neutron stars

The orbital velocities of close binary systems are typically several hundred kilometres per second. (For example, the X-ray source Cen X-3, with a period 2.07 days, has an orbital velocity of 500 km s^{-1}.) When such a system is disrupted by a supernova explosion the components will fly apart, usually with a large proportion of the orbital velocity. The observed velocities of the pulsars may be explained adequately by this process; we might therefore deduce that all high-velocity pulsars come from binary systems. However, there is also a good possibility that a neutron star might leave a solitary supernova with high velocity, as may be seen from the following discussion.

When Shklovsky (1971) pointed out that the proportion of mass of a Type II supernova which ultimately condensed into a neutron star might be much less than one-tenth, he also noted that any asymmetry in the explosion would tend to give a large velocity to the star. Most of the ejected material leaves the system with a velocity of the order of 10^4 km s^{-1}. Only a small asymmetry would be needed to give the neutron star a typical velocity of 100 km s^{-1}; indeed, it was to be remarked that the comparatively *low* velocities of the pulsars suggested that the explosions were notably symmetrical.

The process of an asymmetric collapse was examined in more detail by Michel (1970). The collapsing core is Rayleigh–Taylor unstable during the implosion, so that it is expected to form several condensations of mass. These objects would orbit round one another in a way determined by the original angular momentum of the supernova core, except for the subsequent explosion which drives out a large part of the total mass. The system is then very like a disrupting binary, and the separate objects would leave the system with velocities approaching their orbital velocities. These 'runaway' velocities would typically be in excess of 10^3 km s^{-1}, and might reach 10^4 km s^{-1}.

We conclude that if pulsars are indeed remnants of Type II supernova explosions, whether the original star was solitary or binary, there is no difficulty in explaining either the lack of binary pulsars or the high velocities of most individual pulsars.

19.6 Associations between pulsars and supernovae

Only four out of more than a hundred known pulsars appear to be associated directly with supernova remnants. Of these four only one association (the Crab Pulsar) is certain; another (the Vela Pulsar) is probable; a third (PSR 0611+22 with IC 443) is fairly likely; while the fourth (PSR 1145−62) is a mere possibility. Searching in the other

References

direction, and looking for pulsars within or close to the hundred or more supernova remnants, has not so far yielded any other associations.

The essential parameter in this discussion is the age of the pulsar. If this considerably exceeds the lifetime of the visible supernova remnant, which is of the order of 10^5 years, then no association can be expected. If the supernova remnant is no longer expanding, having encountered a sufficient mass of interstellar material to slow it down from an expansion velocity of say 500 km s^{-1} to 10 km s^{-1}, then the pulsar velocity may have taken it outside the nebula. The observed associations confirm this view. The Crab Pulsar (age 10^3 years) is near the centre of the Crab Nebula. The Vela Pulsar (age 10^4 years) is within a less well defined supernova remnant. The IC 443 nebula (age 10^5 years) has slowed down completely; PSR 1145−62 has a similar age and is found outside the nebula. There is little doubt that most other pulsars have ages in excess of 10^6 years, so that not only have they left the neighbourhood of the supernova explosion; the remnant itself will have disappeared from view.

References

Batten, A. H. (1967). *Ann. Rev. Astron. Astrophys.* **5**, 25.
Blaauw, A. (1961). *Bull. astron. Inst. Neth.* **15**, 265.
Boersma, J. (1961). *Bull. astron. Inst. Neth.* **15**, 291.
Cameron, A. G. W. & Canuto, V. (1973). *Proc. 16th Solvay Conf. on Physics, Brussels*, p. 268.
Colgate, S. A. (1970). *Nature, Lond.* **225**, 247.
Fowler, W. H. & Hoyle, F. (1964). *Astrophys. J. Suppl.* **9**, 201.
Hoyle, F., Fowler, W. H., Burbidge, G. R. & Burbidge, M. (1964). *Astrophys. J.* **139**, 909.
Ilovaisky, S. A. & Lequeux, J. (1972). *Astron. Astrophys.* **18**, 169.
Katgert, P. & Oort, J. H. (1967). *Bull. astron. Inst. Neth.* **19**, 239.
McCluskey, G. E. & Kondo, Y. (1971). *Astrophys. Space Sci.* **10**, 464.
Michel, C. (1970). *Nature, Lond.* **228**, 1073.
Minkowski, R. (1964). *Ann. Rev. Astron. Astrophys.* **2**, 247.
Shklovsky, I. S. (1968). *Supernovae* (New York: John Wiley).
Shklovsky, I. S. (1971). *Astrophys. Lett.* **8**, 101.
van den Heuvel, E. P. J. (1973). *Proc. 16th Solvay Conf. on Physics, Brussels*, p. 119.
van den Heuvel, E. P. J. & De Loore, C. (1973). *Astron. Astrophys.* **25**, 387.
Woltger, L. (1972). *Ann. Rev. Astron. Astrophys.* **10**, 129.
Zwicky, F. (1962). *Problems of Extragalactic Research*, ed. G. C. McVittie, p. 347. (New York: Macmillan).

20
The distribution and the ages of pulsars

A glance through a catalogue of the known pulsars, such as that in the Appendix to this book, shows at once that they are largely found in the Milky Way; they are therefore young galactic objects, and the association with Type II supernovae seems established without further argument. However, the catalogue should be read in conjunction with a description of the surveys which found the pulsars; it is equally obvious that these surveys have mostly concentrated on the plane of the Galaxy, presumably on the basis of a belief based on rather inadequate statistical evidence that the Milky Way is a good place to look for pulsars anyway. So, as with any distribution problem, selection effects must be looked at very carefully.

As will be seen from the descriptions in Chapter 2, a search for pulsars is a multidimensional search, in which the possibility of detection depends not only on the mean flux density of the pulsar but also to some extent on its period and its dispersion measure. Further, the minimum flux density for detection depends on position in the sky: if it is situated in a region of high radio brightness the noise level of the radio telescope is increased and detection becomes more difficult. There are only two substantial surveys in which these factors have been carefully assessed so that the data can be used to determine the distribution of pulsars through the Galaxy.

The first survey to obtain a homogeneous sample was that discussed by Large (1971). This survey covered 7 steradians of the southern sky, producing twenty-nine pulsars with measured flux densities, positions, periods and dispersion measures. It was this survey which first established the strong concentration of pulsars in the galactic plane. The second survey covered a smaller solid angle (1 steradian) with a greater sensitivity. This was the survey at low galactic latitudes (mainly $|b| < 10°$), and covering rather more than one quadrant of the galactic plane, made at Jodrell Bank by J. G. Davies, A. G. Lyne & J. H. Seiradakis (unpublished data). A similar method of analysis was used for both surveys; since the latter survey contained a larger homogeneous sample, including fifty-two pulsars, we shall quote results from this survey, referring only briefly to the earlier analysis by Large.

20.1 The observed and actual distribution functions

If we assume that the distribution of interstellar electrons is known, then the dispersion measure (DM) at once gives a distance for each pulsar. The distribution is not in fact known in detail, but it does seem that a simple model, either of uniform distribution or preferably with an exponential decrease with distance $|z|$ from the plane is sufficiently good for statistical purposes (Chapter 12). From the survey observations we can then tabulate luminosity L, period P and the position of each pulsar. With a comparatively small sample of pulsars, we cannot expect much detail in the galactic distribution, and the position of a pulsar may be specified simply as a radial distance R from the galactic centre, and a distance $|z|$ from the plane. The required distribution is therefore the function $\rho(L, P, R, z)$.

The observed function $N(L, P, R, z)$ differs from ρ because of the sensitivity limits of the survey. For example, the survey technique may involve a reduced sensitivity for lower values of P; the sensitivity may also vary according to the density of sampled areas over the sky and according to the background radio brightness. The result is that the distribution $\rho(L, P, R, z)$ is very poorly sampled. Mathematically we can consider a four-dimensional space containing unit cells on which we place statistical weights W corresponding to the number of samples of each unit made in any particular survey. A test of any assumed $\rho(L, P, R, z)$ can then be found by forming the products:

$$N(L, P, R, z) = W(L, P, R, z)\rho(L, P, R, z) \qquad (20.1)$$

and comparing the computed values of N with the observed numbers.

Obviously this procedure cannot yield adequate results on the basis of only fifty-two pulsars. However, if we assume that the density can be written as the product of four independent functions

$$\rho(L, P, R, z) = \rho_1(L)\rho_2(P)\rho_3(R)\rho_4(z) \qquad (20.2)$$

we find that there is sufficient information to obtain useful descriptions of the four population functions. Of course, we must justify this step by showing that any correlation which may exist between the separated functions does not invalidate the conclusions. This will be examined later.

The procedure is now to assume functions for the four distributions, derive the expected $N(L, P, R, z)$, and modify independently the separate distribution so as to obtain the best agreement with observation. At the same time the accuracy of the derived populations may be tested by noting the sensitivity of the agreement to variations in the population functions.

Distribution and ages of pulsars

The population functions obtained by Davies *et al.* are shown in Figs. 20.1 and 20.2. Each function includes error bars obtained from the variation analysis, so that for example it can be seen that the period function $\rho(P)$ is well determined between $\frac{1}{4}$ and 2 s, and rather poorly at shorter periods where the search procedures tend to be less sensitive.

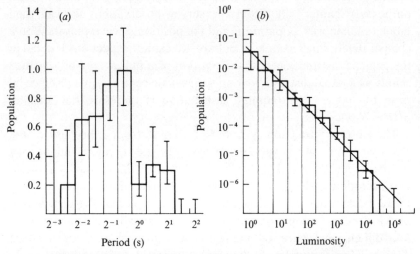

Fig. 20.1. (*a*) Distribution in period. (*b*) Distribution in luminosity. The scales in the histograms of (*a*) and (*b*) depend on the mean electron density $\bar{n}_e$. If $\bar{n}_e = 0.025$, the population in pc^{-3} is obtained by multiplying by $(0.025)^3$, and the luminosity in W Hz^{-1} by multiplying by $(0.025)^{-2}$.

20.2 Distribution in period

The pulsar catalogue shows a clear peak at about $\frac{1}{2}$ s period. Fig. 20.1(*a*) shows that the true distribution, which allows for selection factors, shows the same peak, even though the statistical weights at small periods are poor. We shall later compare this distribution with a theoretical model in which pulsars originally all have very short periods, and evolve towards longer periods: the model fits the observations only at periods up to about $\frac{1}{2}$ s, where we must postulate some cut-off mechanism.

20.3 Distribution in luminosity

The function in Fig. 20.1(*b*) is remarkably simple. Over four decades the population fits the empirical law

$$N(L)\,dL \propto L^{-2}\,dL. \tag{20.3}$$

20.5 Distribution with radial distance R

The units of luminosity depend on the conversion of dispersion measure into distance: they are quoted here as $Jy(DM)^2$, which refers to luminosity at the observed radio frequency of 408 MHz. A conversion to total radiated power depends, of course, on the spectrum (see Chapter 8). If a uniform electron density $n_e = 0.025$ cm^{-3} is adopted (see Chapter 12), then one unit of dispersion measure corresponds to 40 pc.

20.4 Distribution in z-distance

Again the distance scale in Fig. 20.2(a) is quoted in dispersion measure units. Adopting $n_e = 0.025$ cm^{-3} we find that the pulsars extend to a z-distance of 600 pc; if the electron density falls with z-distance then the pulsar distribution must extend further from the plane. Taking this into account (see Chapter 12), the best estimate of the scale height of the pulsar distributions is about 300 pc. This is conspicuously large compared with the scale height of the stellar Population II, from which we suppose the pulsars to have been derived via supernova explosions. The scale height does not seem to vary with radial distance R.

20.5 Distribution with radial distance R

Figure 20.2(b) shows this distribution, with distances calculated assuming that the galactic centre is at a distance corresponding to $DM = 250$ pc cm^{-3}. It is very noticeable in the raw data of the surveys that few

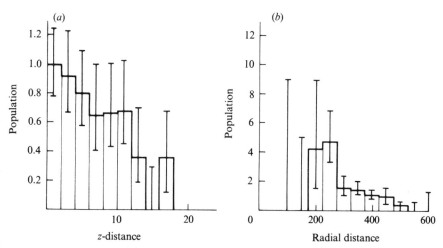

Fig. 20.2. (a) Distribution in z-distance. (b) Distribution in radial distance R. The scales in the histograms of (a) and (b) depend on the mean electron density $\bar{n}_e$. If $\bar{n}_e = 0.025$, the population in pc^{-3} is obtained by multiplying by $(0.025)^3$. The distances are obtained by multiplying by 0.025.

Distribution and ages of pulsars

pulsars are found at longitudes greater than 90° from the centre of the Galaxy; this is seen in the distribution as a concentration of pulsars within a radial distance of 10 kpc, which is taken to be the position of the Sun. This concentration agrees well with the expected distribution of supernova explosions.

20.6 The total galactic population

The right-hand end of Fig. 20.1(b) represents only a small population of intense pulsars: an extrapolation to a luminosity such that only one such pulsar would be expected in the whole Galaxy would only add some tens of pulsars to the known list. These represent a few intense pulsars situated at distances of, say, 10 kpc or more. At the low luminosity end there appears to be a large population of low-luminosity pulsars remaining to be discovered, since it seems unlikely that the monotonic increase in $\rho(L)$ would be found to cut off suddenly if the sensitivity limits of the searches were improved by another factor of ten.

Leaving aside these undiscovered pulsars, we can now integrate the distributions of observable pulsars to find the total population in the Galaxy. This total is 10^5, within a factor of 2 either way.

This large total takes no account of the probable effect of beaming, in which a pulsar may only be observed if the observer lies within a certain range of angles from its rotation axis. On the relativistic beaming theory, this means that the actual population is three times larger, while on the polar cap theory the population must be between ten and twenty times larger. As we shall see, the population is already uncomfortably large for explanation via supernovae, so we shall take the lower factor and give the total population as a minimum of 3×10^5. This figure, combined with the average age of pulsars, will give a rate of formation of pulsars, which can be compared with the rate of supernova explosions.

20.7 Correlations between the population functions

It would not be surprising to find significant correlations between any pair of the four population functions which were treated separately in the previous section. In fact, there is no correlation large enough to upset the conclusions, which were based on an assumption that the functions were completely separable. There are, however, some significant correlations involving the period and its derivative $\dot{P}$ which are of interest in a discussion of the life history of pulsars.

There is a clear correlation between luminosity and $P/\dot{P}$, which is the apparent age. The luminosity falls with increasing age. The correlation coefficient is -0.5 ± 0.1. This effect is not sufficient in itself to account for

20.8 Ages of the pulsars

the fall in population for periods greater than $\frac{1}{2}$ s, which must be a more catastrophic cut-off of emission.

The second correlation occurs as a relation between z-distance and $P/\dot{P}$. One might expect the pulsars to be formed close to the plane, and move away from it with a high velocity: the consequence would be a positive correlation between age and z-distance. In fact the correlation is very poor. Only for the youngest pulsars is there any correlation. Dividing the pulsars into two groups, with $P/\dot{P}$ above and below 10^6 years, gives a significant correlation for the 'young' pulsars and none for the 'old' pulsars, which are the majority.

20.8 The ages of the pulsars

The simple theory of the slowdown of a pulsar, as set out in Chapter 7, assumes that the dissipation of rotational energy changes the angular velocity according to the law

$$\dot{\Omega} = -k\Omega^n. \tag{20.4}$$

The age of a pulsar, from a time at which the period was very short until it reached a considerably longer period P was then given by $(1/n - 1)(P/\dot{P})$. Usually the energy is considered to be lost by magnetic dipole radiation, when k is proportional to the square of the dipole moment and $n = 3$; the age is then $\frac{1}{2}P/\dot{P}$.

If all pulsars were identical at birth, such a law would lead to a unique relation between P and $\dot{P}$, such that a logarithmic plot would show a straight line with slope -1. In practice the plot shows a remarkably wide scatter (Fig. 20.3); the formal correlation coefficient for this group of eighty pulsars is zero. Obviously in this plot each point is pursuing a track sloping downwards in the direction of increasing P and decreasing $\dot{P}$, but the scatter of points indicates that the individual tracks are very different. It has been suggested by Lyne, Ritchings & Smith (1975) that the distribution of points on this P, $\dot{P}$ plot is arrived at through curved evolutionary tracks, in which the factor k in (20.4) decreases exponentially with time. This would mean that $\frac{1}{2}P/\dot{P}$ no longer represents age.

We have already seen in Chapter 7 that the known proper motions of pulsars show that they left the galactic plane at epochs which are later than those suggested by their apparent 'ages' $\frac{1}{2}P/\dot{P}$. There are features of the P, $\dot{P}$ diagram (Fig. 20.3) which also suggest a discrepancy with the simple evolutionary theory: in particular there are no pulsars in the lower left of the diagram representing pulsars evolving along a slope of -1 towards the points with small $\dot{P}$. Lyne et al. (1975) therefore suggest that

Distribution and ages of pulsars

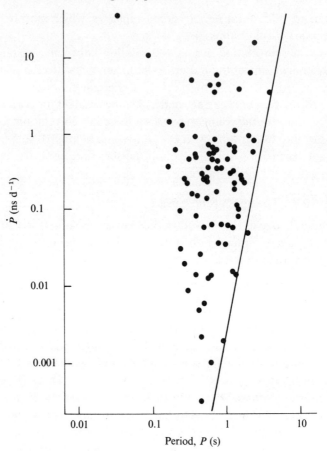

Fig. 20.3. The relation between period P and its derivative $\dot{P}$. The sloping line is the cut-off line at which emission ceases.

the period could evolve according to

$$P\dot{P} = K_0 \exp(-2t/\tau_D)$$

where τ_D is a decay time constant for the magnetic dipole moment. The effect is seen in Fig. 20.4 which shows the evolutionary tracks for $\tau_D = 10^6$ years and for a 10:1 range in initial strength of dipole moment (i.e. 10^2:1 in value of K_0). The actual ages of the pulsars are shown on broken lines crossing the evolutionary tracks.

An evolution of this kind is not the inevitable consequence of the form of the $P, \dot{P}$ distribution; it is so far only the simplest explanation of this form and of the age discrepancy as seen in the proper motions.

20.9 The radiation cut-off

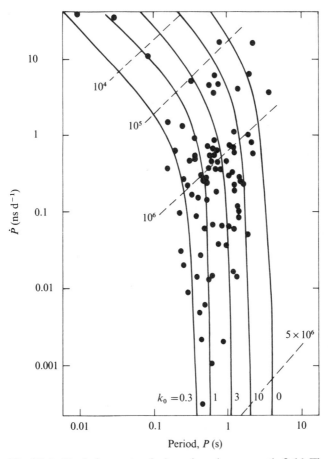

Fig. 20.4. Evolutionary tracks for a decaying magnetic field. The ages of the pulsars in years are indicated by the broken line.

20.9 The radiation cut-off

The form of the $P, \dot{P}$ distribution also provides a description of the conditions under which a pulsar stops radiating strong radio pulses. Fig. 20.3 shows that the observed distribution is confined to the left of a line which may be called a cut-off line. The spin slowdown must take pulsars across this line, and they must stop emitting strong radio pulses at about this time, since the lack of pulsars to the right of the line is not due to observational selection.

The slope of the cut-off line is approximately $+\frac{1}{5}$, so that the condition for the cut-off seems to depend on the product $\dot{P}P^{-5}$. It is interesting to note that this is directly related to the strength of the magnetic field at the

Distribution and ages of pulsars

velocity-of-light cylinder; we know that the dipole moment varies as $(P\dot P)^{1/2}$ from the slowdown theory (Chapter 7), so that if the field falls as the cube of the radial distance it is proportional to $\dot P^{1/2} P^{-5/2}$ at the velocity-of-light cylinder. This provides supporting evidence for theories in which the origin of the radio pulses is close to this radial distance rather than close to the surface.

It is interesting to note also that those pulsars which show a characteristic pulse drifting behaviour (Chapter 9) tend to be close to the cut-off line. They may therefore be regarded as the older pulsars, approaching the cut-off condition. The same appears to be true of pulsars showing pulse nulling, so that nulling may represent the last spasmodic flickers of radiation. PSR 1944+17 is a striking example: it is very close to the cut-off line and with a very low value of $\dot P$, and it characteristically radiates for several periods followed by a similar or longer completely quiet interval. The degree of complexity in the integrated profile may also be an indicator of old age, as suggested by Huguenin, Manchester & Taylor (1971), although there may be some observational selection effects at play in the determination of the complexity.

20.10 Conclusions

We have seen that several characteristics of the pulsars suggest strongly that their ages cannot be measured as $\tfrac{1}{2} P/\dot P$, which often indicates a much larger age due to the decay of the magnetic dipole moment, which reduces the slowdown rate $\dot P$. The extreme 'age' $\tfrac{1}{2} P/\dot P$ for some pulsars reaches 10^8 or 10^9 years, while the actual age for these pulsars need be no more than 10^6 to 10^7 years. This age is close to τ_D, the decay time constant of the dipole moment; in general, if the conventional 'age' is given by $T = (1/n - 1)(P/\dot P)$, the true age t is given by

$$t = \tfrac{1}{2}\tau_D \log\left(\frac{2T}{\tau_D} + 1\right).$$

The cut-off of radio emission, which appears to be a catastrophic cut-off, as if it suddenly becomes impossible for the pulsar to sustain a coherent oscillation, occurs when the magnetic field at the velocity-of-light cylinder falls below a critical value.

We therefore take the minimum population of observable pulsars in the Galaxy as 3×10^5, and the maximum lifetime for most pulsars as 3×10^6 years. It is an easy calculation to find that one pulsar must be created every 10 years to sustain such a population, a rate which is five times greater than the presently estimated rate of supernova explosions. Furthermore, there may in fact be many more pulsars, since the luminos-

References

ity function suggests that more exist at low luminosity, and in addition the beaming factor may have been underestimated in our present estimate of the total number of pulsars. Again, it would be surprising if every supernova explosion produced a neutron star, and that all neutron stars were observable as strong pulsating radio sources.

There seems, therefore, to be a serious discrepancy between the theory of origin of pulsars in supernovae, and the observations of their ages and numbers in the Galaxy. No other source for pulsars has, however, been postulated, and the discrepancy remains unresolved.

References

Huguenin, G. R., Manchester, R. M. & Taylor, J. H. (1971). *Astrophys. J.* **169**, 97.
Large, M. I. (1971). *IAU Symposium No. 46*, p. 165. (Dordrecht: D. Reidel.)
Lyne, A. G., Ritchings, R. T. & Smith, F. G. (1975). *Mon. Not. R. astron. Soc.* **171**, 579.

21
High energies and condensed stars

At the end of this account of the pulsars, we should recall the state of astrophysics at the time of their discovery. At that time the term 'black hole' had not entered the astrophysical vocabulary; now it is in common usage. The concept of condensed stars reached only to white dwarfs, except in the minds of a very few theorists. The certainty that pulsars were the hitherto undiscovered neutron stars came slowly, for the concept of such a degree of condensation was hard to accept. General acceptance of the identification has, as we have seen, left unsolved problems in the structure of neutron stars and even more in their magnetospheres. The concept of condensed stars as a source of energetic processes has, however, found other applications, which form the subject of this final chapter.

21.1 X-ray binaries

Developments in X-ray astronomy are so rapid at the time of writing that the account in Chapter 4 is impossible to keep up to date. At a conference in Leicester (August 1975: *Nature, Lond.* **257**, 275) the first results from the satellite Ariel V were presented. By a fortunate coincidence this satellite had just observed an X-ray nova, a new source which became within a few days the brightest X-ray source in the sky. This object (A0621−00) is probably very similar to several other transient sources which are now known; it is estimated that a new one is observable every 10 days or so. The effect on the catalogues is chaotic; early in its life the satellite Ariel V recorded the loss of eight sources which faded from the Uhuru catalogue, and the gain of twenty-four new sources, which probably had become brighter since the Uhuru survey a few years earlier.

These transient sources are thought to be binaries, in which the streaming of matter on to a neutron star or black hole is either variable or occasionally hidden by an occultation. If this is so, then these X-ray sources are closely related to the pulsars, representing the remains of a supernova explosion in a binary system which remained bound. Such systems might not have as high a velocity as the pulsars, and we might expect to see X-ray sources close to the supernova remains in which they

21.2 X-ray in clusters

originated. There are X-ray sources close to the remnants S147 and G321.9−0.3 (D. H. Clark & J. H. Parkinson, at the Leicester Conference), and it will be interesting to explore these possible associations by estimating the ages and velocities of the sources.

Periodicities of the order of some minutes have been found in several X-ray sources, e.g. A1118−61 (6.75 minutes), A0535+26 (7 minutes), A0900−40 (4.7 minutes). These are very long for rotation times of neutron stars: a normal pulsar would scarcely reach such a slow rotation in the age of the Universe at the observed slowdown rates. The period might be the orbital period of a very close binary, although this would imply a very short lifetime through gravitational radiation. More likely, the period is indeed a slow rotation, resulting from an unusually efficient braking mechanism which applies only to binary systems, and not to solitary pulsars. This mechanism is the effect of mass transfer on to the neutron star. If there is a developed magnetosphere round the star, the infall of mass can develop a large drag on the star through the magnetic coupling between the magnetosphere and the star.

These physical processes should become clearer as the observations give more detail of the 'light curves' of the periodic variations. It may then be possible to estimate the ages of the X-ray sources, and their population in the Galaxy, allowing a comparison to be made with the evolutionary theory presented in the previous chapter. Meanwhile we accept that most X-ray sources comprise neutron stars in binary systems, differing from pulsars basically in the existence of mass transfer.

21.2 X-ray in clusters

About 4% of the X-ray sources are identified with Globular Clusters in the Galaxy. They are intrinsically powerful sources, and they are about a hundred times more powerful than might be expected from the numbers of stars in the clusters, assuming that those stars are as likely to be X-ray sources as the average star in the Galaxy. Alternatively expressed, four out of 100 X-ray sources are in clusters, but only one star in 10^4 is in a cluster. A new mechanism for radiation is involved, which can produce about 10^{39} to 10^{40} erg s^{-1}, and which is found only in the high concentration of stars at the centre of a cluster.

Two suggestions are under discussion. First, there may be such a mass concentration that a single massive body at the centre becomes a black hole, with matter falling in from the rest of the cluster. Second, the dynamics of the cluster centre might involve such frequent near-collisions that mass interchange becomes practically continuous, so that the central region is a complex of many interacting stars. In this second theory, close

High energies and condensed stars

encounters between individual stars may dissipate kinetic energy through tidal forces, creating an unusually large number of binary systems which together form the X-ray source.

Resolution of these opposing theories may become possible through a measurement of the angular diameters of the cluster X-ray sources.

21.3 The galactic centre

The radio source Sag A, close to or at the centre of the Galaxy, is known to contain an intense and very small component, less than 0.005 pc in diameter. Associated with the radio source is a strong emitter of infrared radiation. One of the transient X-ray sources (A1742−28) discovered by Ariel V has now been identified with Sag A (Eyles, Skinner, Willmore & Rosenberg, 1975). The intensity of the X-ray source doubled within 2 days, indicating that the emitter must be small, probably less than 0.002 pc in diameter.

The galactic centre is evidently much more important than a mere geometric construction of the centre of rotation. It is known from 21-cm hydrogen line studies to contain a very high mass concentration. Close to it there are gas clouds whose velocities contain a large outward radial component, either because they are in peculiar elliptical orbits or more probably because they have come from the galactic centre after a series of massive explosions. The nature of these explosions, and of the centre itself, are unknown, but it seems likely that within the large concentration of mass there will be either a condensed star of very large mass, forming a black hole, or a large number of neutron stars.

21.4 Extragalactic clusters

X-ray sources have now been discovered in several extragalactic clusters, and notably the Virgo, Coma and Perseus clusters. This new type of X-ray source is stronger than the sum of the X-ray emission expected from the cluster galaxies, assuming them to be similar to our own Galaxy. Two possible explanations place the emission either in hot gas pervading the whole clusters, or in some especially bright individual galaxies. These two theories may be distinguishable in future observations with higher angular resolution. There is naturally a speculation of a hierarchical structure of X-ray sources, with the source at our galactic centre forming a weak counterpart to a massive source at the centre of the cluster of galaxies. Although such a source would help solve the problem of the so-called 'missing mass', indicated by dynamical studies of clusters, its existence can only be a matter of speculation at this time.

21.5 Radio galaxies and quasars

Finally we recall that the problem of energy generation in the extragalactic radio sources remains unsolved. A typical radio source emits up to 10^{48} erg s^{-1} over a period of 10^5 or 10^6 years, with a total output exceeding 10^{60} erg. Even at the moment of observation it must contain over 10^{60} erg in the form of high-energy particles and magnetic field, so that it is most improbable for the origin to lie in a supernova type of explosion. It is much more reasonable to look for an energy release due to a gravitational collapse; furthermore this energy release must be mediated through a rotation of the collapsing galaxy. Evidence for this comes from the common existence of a persistent and well-defined axis in quasars, which commonly comprise a point-like centre with two outlying regions of radio emission, forming a straight line. These outer radio components appear to be expanding outwards, and in many quasars and radio galaxies there is evidence of other components following outwards on the same path.

The simplest interpretation is that a galaxy collapsing under its own gravitational attraction is converting gravitational potential energy into energy of rotation and thence into high-energy particles. The mechanism produces particles only near the rotational poles, streaming outwards in two diametrically opposite directions. Whether the energy release in quasars can be understood in terms of pulsar theory is an open question, but it would be gratifying if the strange behaviour of the particles near neutron stars could eventually be used as a key to one of the most energetic and most mysterious processes in astrophysics.

Reference

Eyles, C. J., Skinner, G. K., Willmore, A. P. & Rosenberg, F. D. (1975). *Nature, Lond.* **257**, 291.

Appendix. The positions and periods of 105 pulsars

PSR	Galactic co-ordinates		Period (s)	$\dot{P}$ (ns d^{-1})	DM (pc cm^{-3})
	l	b			
0031−07	110.4	−69.8	0.943	0.0359	10.89
0105+65	124.6	3.3	1.284	1.074	30
0138+59	129.1	−2.3	1.223	0.015	34.5
0153+61	130.5	0.2	2.352	16.358	60
0254−54	270.9	−54.9	0.448	—	10±5
0301+19	161.1	−33.2	1.386	0.103	15.7
0329+54	145.0	−1.2	0.715	0.177	26.78
0355+54	148.1	0.9	0.156	0.379	57.0
0450−18	217.1	−34.1	0.549	0.360	39.9
0525+21	183.8	−6.9	3.745	3.462	50.8
0531+21	184.6	−5.8	0.033	36.526	56.8
0540+23	184.4	−3.3	0.246	1.334	72
0611+22	188.7	2.4	0.335	5.161	96.7
0628−28	237.0	−16.8	1.224	0.217	34.36
0736−40	254.2	−9.2	0.375	0.086	100±10
0740−28	243.8	−2.4	0.167	1.453	80±15
0809+74	140.0	31.6	1.292	0.014	5.77
0818−13	235.9	12.6	1.238	0.182	40.9
0823+26	197.0	31.7	0.531	0.145	19.4
0835−45	263.6	−2.8	0.089	10.823	69.2
0834+06	219.7	26.3	1.274	0.587	12.9
0835−41	260.9	−0.3	0.767	—	120±12
0904+77	135.3	33.7	1.579	—	—
0940−55	278.5	−2.1	0.664	—	145±15
0943+10	225.4	43.2	1.098	0.305	15.35
0950+08	228.9	43.7	0.253	0.020	2.97
0959−54	280.1	0.3	1.437	—	90±10
1055−51	286.0	7.0	0.197	—	<30
1112+50	155.1	60.7	1.656	0.224	9.2
1133+16	241.9	69.2	1.188	0.323	4.83
1154−62	296.7	−0.2	0.401	—	270±35
1221−63	300.0	−1.3	0.216	—	92
1237+25	252.2	86.5	1.382	0.083	9.25
1240−64	302.1	−1.6	0.389	—	220±20

Appendix

PSR	Galactic co-ordinates		Period (s)	$\dot{P}$ (ns d^{-1})	DM (pc cm^{-3})
	l	b			
1323−62	307.1	0.3	0.530	—	313
1354−62	310.5	−0.6	0.456	—	400
1359−50	314.5	11.0	0.690	—	20±10
1426−66	312.3	−6.3	0.787	—	60±6
1449−65	315.3	−5.3	0.180	—	90±10
1451−68	313.9	−8.6	0.263	0.259	8.6
1508+55	91.3	52.3	0.740	0.435	19.60
1530−53	325.7	1.9	1.369	—	20±5
1541+09	17.8	45.8	0.748	0.038	35.0
1556−44	334.5	6.4	0.257	—	—
1557−50	330.7	1.6	0.193	—	270
1558−50	330.7	1.3	0.864	—	165
1601−52	329.7	−0.6	0.658	−	35
1604−00	10.7	35.5	0.422	0.026	10.72
1641−45	339.2	−0.2	0.455	—	449
1642−03	14.1	26.1	0.388	0.154	35.71
1700−32	351.7	5.4	1.212	0.059	103
1700−18	4.0	14.0	0.802	—	<40
1706−16	5.8	13.7	0.653	0.550	24.9
1717−29	356.5	4.3	0.620	—	45
1718−32	354.5	2.5	0.477	0.065	120
1727−47	342.6	−7.6	0.830	—	121±4
1730−22	4.0	5.8	0.872	0.002	45
1742−30	358.5	−1.0	0.367	0.921	84
1747−46	345.0	−10.2	0.742	6.040	40±10
1749−28	1.5	−1.0	0.563	0.711	50.9
1813−26	5.2	−4.8	0.593	0.001	90
1818−04	25.5	4.7	0.598	0.546	84.5
1819−22	9.4	−4.3	1.874	0.050	140
1822−09	21.4	1.3	0.769	4.512	19.3
1826−17	14.6	−3.3	0.307	0.483	207
1831−03	27.7	2.3	0.687	3.582	235
1831−04	27.0	1.8	0.290	0.009	68
1845−01	31.3	0.2	0.659	0.453	90
1845−04	28.9	−1.0	0.598	4.480	141.9
1846−06	26.7	−2.4	1.451	3.949	152
1857−26	10.5	−13.5	0.612	0.014	35±10
1858+03	37.2	−0.6	0.655	0.646	402±2
1900−06	28.5	−5.6	0.432	0.298	180
1900+01	35.8	−1.9	0.729	0.356	228
1906+00	35.1	−3.9	1.017	0.467	111
1907+02	37.7	−2.7	0.495	0.248	190
1907+10	44.8	1.0	0.284	0.232	144

Appendix

PSR	Galactic co-ordinates		Period (s)	$\dot{P}$ (ns d^{-1})	DM (pc cm^{-3})
	l	b			
1910+20	54.0	5.0	2.233	0.825	84
1911−04	31.3	−7.1	0.826	0.351	89.4
1915+13	48.3	0.6	0.195	0.621	94
1917+00	36.5	−6.1	1.272	0.645	85
1918+19	53.9	2.7	0.821	0.065	140
1919+21	55.8	3.5	1.337	0.116	12.43
1920+10	55.3	3.0	1.078	0.706	220
1929+10	47.4	−3.9	0.227	0.100	3.18
1933+16	52.4	−2.1	0.359	0.519	158.5
1944+17	55.3	−3.5	0.441	0.002	16.3
1946+35	70.6	5.0	0.717	0.607	129.1
1953+29	66.0	0.7	0.427	0.003	20
2002+30	67.7	−0.7	2.111	6.416	233
2016+28	68.1	−4.0	0.558	0.013	14.18
2020+28	68.9	−4.7	0.343	0.160	24.6
2021+51	87.9	8.4	0.529	0.264	22.58
2045−16	30.5	−33.1	1.962	0.947	11.51
2106+44	86.9	−2.0	0.414	0.005	129
2111+46	89.0	−1.3	1.015	0.062	141.4
2148+63	104.1	7.4	0.380	0.014	125
2154+40	90.5	−11.4	1.525	0.283	71.0
2217+47	98.4	−7.6	0.538	0.239	43.54
2223+65	108.7	7.0	0.683	0.825	—
2255+58	108.8	−0.7	0.368	0.497	148
2303+30	97.7	−26.7	1.576	0.251	49.9
2305+55	108.6	−4.2	0.475	0.006	45
2319+60	112.0	−0.6	2.256	0.588	96±3
2324+60	112.9	0.0	0.234	0.031	120

Index

age of pulsars
  apparent, due to proper motion, 70
  compared with supernova remnants, 219
  Crab Pulsar, 61
  from period change, 59, 68
  related to magnetic decay, 225, 228
  related to other characteristics, 184
angular diameter, due to scintillation, 149, 165–6
Ariel V satellite, 230

Baade's star, 113, 117
bandwidth
  of pulsar radiation, 182, 206
  of receiver, 11
barycentre of Solar System, 57
beaming mechanisms, 186 et seq.
Bell, Jocelyn, 2, 3
binary star systems
  mass transfer, 216
  PSR 1613+16, 73
  pulsars in, 60, 72, 217
  X-ray sources, 25 et seq., 32, 230
black hole, 43, 217, 230–1
  Cyg X-1, 32-3
braking index, 69, 114
Burnell, Mrs J., see Bell, Jocelyn

Cen X-3, 29
Cepheid variables, 19
circular polarisation
  cyclotron radiation, 172, 175
  individual pulses, 95
  integrated profiles, 83, 85
  synchrotron radiation, 174
clusters as X-ray sources, 231
coherence in pulsar radiation
  discussion, 208–9
  high intensity, 182
  particle bunching, 178, 207
core of neutron star, 40, 68
co-rotation of magnetosphere, 49
cosmic rays
  effect on interstellar gas, 130, 156
  generated by pulsars, 47
Crab Nebula, 109 et seq.

Crab Pulsar, 117 et seq.
  discovery of optical pulses, 5, 14
  discovery of radio pulses, 6
  giant pulses, 91, 123
  glitch, 61, 66
  irregularities in timing, 61–4
  period increase, 6
  polarisation: optical, 125; radio, 126
  proper motion, 70
  pulse broadening, 163–4
  pulse profiles, 120
  spectrum, 118–19, 205
  timing, 60
  X-ray pulses, 8
critical frequency
  curvature, 177
  synchrotron, 174
crust of neutron star, 67
curvature radiation, 177–8
cut-off in pulsar radiation, 222, 227
cyclotron radiation, 171
Cyg X-1, 32–3

de-dispersion techniques, 13, 14
  applications, 96, 100, 124
degenerate matter, 36, 39
detection, sensitivity of
  optical, 17
  radio, 11
  X-ray and gamma-ray, 17
discovery of pulsars, 1
dispersion measure (DM) (xii), 12
  as distance scale, 129
  at high galactic latitude, 137
  variable, 121
distances of pulsars
  Crab Pulsar, 111, 129
  from dispersion measure, 129
  from hydrogen absorption, 134
Doppler shift in
  binary pulsar, 73
  cyclotron radiation, 173
  relativistic beaming, 207
drifting, 91–3, 103–5
  direction, 105
  related to age 184, 228

electron density
  in Galaxy, 131, 134
  in HII regions, 132
  in magnetosphere, 48
energy density in magnetosphere, 204
energy flow from pulsar, 53–4
equivalent width of pulse profiles, 78
extragalactic sources
  radio, 233
  X-ray, 232

fan beam of emission, 195
Faraday rotation
  Galaxy, 141
  pulsars, 142
  rotation measure (xii), 141 *et seq.*
field strength at emitter, 204
fluctuation spectrum, 106–7
flux density, 11
  spectrum, 89
  variability, 88

Galaxy
  centre, 232
  interstellar medium, 129
  magnetic field, 143–5
  radio emission, 129 *et seq.*, 141
gap in magnetosphere, 52–3
giant pulses, 91, 122–4
glitch
  interpretation, 65–8
  observations, 24, 61, 64–5
glitch function, 65
Gold, T., 5, 23, 45
gravitation at neutron star surface, 47
gravitational radiation, 21, 22
gyrofrequency, 172, 207

Hankin's notch, 85
Her X-1, 26
Hewish, A., xi, 2–4, 55
hydrogen in Galaxy
  ionised, 132–3
  neutral, 130, 134
hyperons, 41

identification of pulsars with neutron stars, 4, 19 *et seq.*
individual radio pulses, 91 *et seq.*
  Crab Pulsar, 123
  polarisation, 93 *et seq.*
  summary of properties, 181–3
integrated pulse profiles, 76 *et seq.*
  Crab Pulsar, 120
  summary of properties, 180
interpulse, 80, 181, 208
interstellar electrons, 129 *et seq.*
inverse Compton radiation, 205

Larmor frequency, 171
location of emitter, 188, 199–203, 228
luminosity distribution, 222

magnetic dipole field
  as energy source, 5, 23, 45, 69
  axisymmetric, 48
  distortion by rotation, 189
  oblique, 51
  secular decay, 70, 226–8
  strength, 45, 184
magnetobremsstrahlung, *see* synchrotron radiation
magnetosphere, 45 *et seq.*
maser amplification, 179, 209
mass of neutron star, 41–3
mass transfer in binaries, 30, 216
microstructure, 91, 100, 182
mode changing, 85–8, 106, 181

neutrinos, 213
neutron drip point, 39
neutron-rich nuclei, 38
neutron stars
  identification of with pulsars, 4, 19 *et seq.*
  magnetosphere, 45 *et seq.*
  structure, 36 *et seq.*; 66
North Galactic Spur, 144
nulling, 106, 181, 228

on/off ratio, 80, 194
oscillations of stars, 19

Pacini, F., 5, 21, 23, 45
pair creation, 51
period of pulsars
  decrease in X-ray pulsars, 30–1
  distribution, 222
  fluctuations, 63, 68
  increase, 23, 59, 68, 183, 225
  measurement, 59
  phase modulation, 64
  X-ray sources, 231
planetary systems, 71
polar cap, 50, 187, 199, 209
polarisation
  circular, 83, 85, 95
  individual pulses, 93
  integrated pulses, 82, 181, 196
  optical: Crab Nebula, 111; Crab Pulsar, 124
  starlight, 140
  Stokes parameters, 82, 195
population of pulsars, 224
position angle of polarisation, 22, 82–3, 124–5
positions of pulsars
  accuracy, 57

determination, 55, 59
pre-discovery records of pulsars
  optical 6,
  radio, 1
  X-ray, 1, 8
proper motion, 70–1
pulse broadening
  diffraction theory, 160
  geometric theory, 158
  observations, 157, 162–4
  variations in, 164
pulse energy histogram, 101–2

quasars, 233

radiation processes, 171 *et seq.*
  coherence, 178, 207
  curvature, 178
  cyclotron, 171
  synchrotron, 174
radio galaxies, 233
radio pulses, individual, *see* individual radio pulses
radius of neutron star, 41–3
recombination lines, 130, 133
relativistic beaming
  discussion, 202
  theory, 190–5
relativity
  in elliptical Earth orbit, 57–8
  in binary pulsar, 74
  rotation measure (RM) (xii), 141 *et seq.*
Ruderman's whiskers, 52

scintillation, 148 *et seq.*
  observations, 2, 153 *et seq.*
  source diameter, 165
  thick screen theory, 152
  thin screen theory, 148
Sco X-1, 26
Solar System
  barycentre, 57
  pulse travel time, 56
spectral index
  observations, 88, 182, 205
  relativistic beaming, 193, 195
spectrum of radio pulses
  individual, 101, 183
  integrated, 88, 182
starlight, polarisation of, 140
starquakes, 67

Stokes parameters, 82
  relativistic transformation, 195
sub-pulses, 91, 181
  drifting, 103 *et seq.*
  polarisation, 96–8, 182, 195
  spectrum, 101
  width, 96, 99, 182, 190
superconductivity, 40, 46
superfluidity, 40
supernovae, 211 *et seq.*
  energy release, 112
  relation to neutron stars, 43, 218
supernova remnants, 214, 219, 231
  Crab Nebula, 109 *et seq.*
surveys
  Jodrell Bank, 15, 220
  Molonglo, 13, 220
  optical, X-ray and gamma-ray techniques, 15–17
  radio techniques, 11 *et seq.*
synchrotron radiation
  Crab Nebula, 111
  Galaxy, 140
  theory, 174 *et seq.*

temperature of neutron star, 36–8.

Uhuru satellite, 26–7, 230

variability, long-term, 88
Vela Pulsar
  discovery, 5, 21
  timing, 64
velocity-of-light cylinder, 49, 188, 228
velocity of pulsars
  from timing, 70
  from scintillations, 166–9
  origin, 218

wind zone, 49
wisps in Crab Nebula, 113, 115

X-ray emission
  Crab Nebula, 113
  Crab Pulsar, 8, 117–18, 203, 205
  thermal, 33–4
X-ray sources
  pulsars, 25 *et seq.*
  others, 230 *et seq.*

z-distance of pulsars, 137–8, 223
Zeeman effect, 141